METAL AND POLYMER MATRIX COMPOSITES

METAL AND POLYMER
MATRIX COMPOSITES

by

Jonathan A. Lee and Donald L. Mykkanen

BDM Corporation
Huntsville, Alabama

WITHDRAWN

NOYES DATA CORPORATION
Park Ridge, New Jersey, U.S.A.
1987

Published in the United States of America by
Noyes Data Corporation
Mill Road, Park Ridge, New Jersey 07656

10 9 8 7 6 5 4 3 2 1

Library of Congress Cataloging-in-Publication Data

Lee, Jonathan A.
    Metal and polymer matrix composites.

    Includes bibliographies and index.
    1. Metallic composites. 2. Polymeric composites.
I. Mykkanen, Donald L. II. Title.
TA481.L44   1987   620.1'18   86-31202
ISBN 0-8155-1111-6

Foreword

This book describes the status of research on metal and polymer matrix composite systems by government and commercial laboratories. It presents data on mechanical, thermal, and physical properties of these advanced metal matrix and polymer matrix systems.

Metal matrix composites (MMCs) have advantages over monolithic metals which include better fatigue resistance, better wear resistance, better elevated temperature properties, higher strength-to-density ratios, higher stiffness-to-density ratios, and lower coefficients of thermal expansion. MMCs also have advantages when compared to polymer matrix composites, including higher functional temperatures, higher transverse strength and stiffness, no moisture absorption, better conductiveness, and better radiation resistance. MMCs also have disadvantages including high costs, newer technologies, complex fabrication methods, and limited service experience.

Polymer matrix composites are highly anisotropic. Strength and stiffness are high parallel to the fibers, but low perpendicular to the fibers. PMCs have stress-strain curves that are generally linear to failure. Polymer matrix materials result in composites that have higher specific tensile strength and stiffness properties. Polymer composites are more advanced in fabrication technology and have lower raw material and fabrication costs.

The information in the book is from *Advanced Materials Research Status and Requirements. Volume I—Technical Summary,* and *Volume II—Material Properties Data Review,* prepared by Jonathan A. Lee and Donald L. Mykkanen of BDM Corporation, for the U.S. Army Strategic Defense Command, March 1986.

The table of contents is organized in such a way as to serve as a subject index and provides easy access to the information contained in the book.

NOTICE

Contents and Subject Index

PART II
MATERIALS PROPERTIES REVIEW

Part I

Technical Summary

The information in Part I is from *Advanced Materials Research Status and Requirements. Volume I—Technical Summary,* prepared by Jonathan A. Lee and Donald L. Mykkanen of BDM Corporation for the U.S. Army Strategic Defense Command, March 1986.

1. Introduction

1.1 <u>Purpose</u>. This document is Volume 1 of a two-volume report describing the status and requirements of the advanced composite materials research in government and commercial laboratories. This task consists of reviewing and evaluating the advanced composite materials which might provide a major step forward in the performance of strategic defense interceptors. This task focused on the application and use of the available and near-term (5 plus years) advanced composite materials. Because of the time limitation, the scope of the technical material goals examined is restricted to advancements in composite materials with polymer and metal matrices. The cost analysis herein is limited to an estimation of the expected raw material costs in the five-year time period. The information contained in this study is the result of a thorough search of the Defense Technical Information Center (DTIC) literature, contractor reports, the Metal Matrix Composites Information Analysis Center (MMCIAC), and open literature. The material examined covers the period from 1975 to mid-1984. Volume II presents data on the mechanical, thermal, and physical properties of general interest advanced metal matrix and plastic (polymer) systems. Because advanced composite materials are in a state of evolution in terms of property improvements, it is not possible to provide final property values in the same sense as those now available for conventional metal alloys. However, Volume II is intended to inform the reader in general terms rather than to serve as a standard sourcebook for the advanced composite systems.

1.2 <u>Applications</u>. This document provides a review of several of the most prominent metal matrix and polymer matrix composite materials. The systems that have been chosen for this study are being seriously considered for engineering structural application to U.S. Army Strategic Defense Command (USASDC) advanced material systems. Figure 1-1 shows the advanced materials examined in this study.

Graphite, boron, Kevlar, silicon carbide, and fiberglass are the principal reinforcement materials considered. Although not truly an advanced reinforcement, fiberglass is included because it is used extensively in

2

military and commercial systems and products.   Aluminum, magnesium, and
titanium are the most important metal matrices.   Epoxy, phenolic, and
polyimide are the most important polymer matrices.

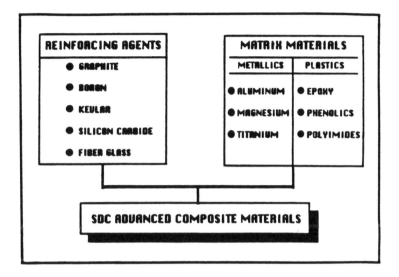

Figure 1-1.   Advanced composite materials selection for
              USASDC material program study.

Figure 1-2 shows how metal matrix composite (MMC) materials may be applied to advanced endoatmospheric interceptor structures. The key advanced endoatmospheric interceptor forebody design requirements include high body bending frequency, minimum body deflections, light weight, and hardness to nuclear and directed energy weapons. The attributes of the MMC materials needed to meet these key design requirements are high specific stiffness and strength at high elevated temperatures, and high thermal and electrical conductivity.

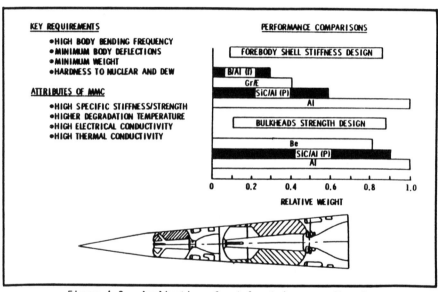

Figure 1-2.   Application of metal matrix composites for advanced endoatmospheric interceptor structures.

Figure 1-3 shows how MMC materials may be applied to advanced exoatmospheric interceptor structures.   Key requirements for exo interceptor structural design include minimum body weight, high body stiffness, hardness to nuclear and directed energy weapons (DEW), and low cost. Potential uses of MMC materials for exoatmospheric interceptor structures can also be found in kill vehicle (KV) external and sensor internal structures.

KEY REQUIREMENTS

- LOW COST
- MINIMUM WEIGHT
- STIFFNESS
- HARDNESS TO NUCLEAR AND DEW

ATTRIBUTES OF MMC

- HIGH SPECIFIC STIFFNESS
- HIGH ELECTRICAL CONDUCTIVITY
- HIGH THERMAL CONDUCTIVITY
- HIGHER DEGRADATION TEMPERATURE

POTENTIAL APPLICATIONS

- KILL VEHICLE (KV) EXTERNAL STRUCTURE
- KV SENSOR INTERNAL STRUCTURES
  (MIRRORS, EMP SHIELDS, SUPPORT)

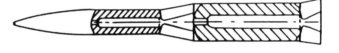

Figure 1-3.   Application of metal matrix composites for advanced exoatmospheric interceptor structures.

2. Material Selection Criteria

The material selection criteria of a composite material system for an advanced interceptor structure are based on the material design requirements and the material selection factors. The material design requirements include the key requirements for interceptor structure design and the material physical properties and characteristics.

2.1 Design Requirements. The key design requirements for advanced interceptor structures are minimum body weight, high body stiffness, and high body strength at elevated temperatures. In addition, the launch and nuclear threat environment survivability constitute a significant factor in structure design requirements. Figure 2-1 summarizes the structural environmental threats.

At any time during a flight, the interceptor may be subjected to blast and radiation loading from a hostile weapon. The interceptor structures may also be subjected to excessive heat loads from thermal radiation and aerodynamic loadings. The interceptor maneuvering loads, inside and outside the atmosphere, provide axial and lateral loads to the structure. Therefore, in selecting candidate materials for use in interceptor support structure, the material design requirements must be carefully evaluated to ensure adequate thermal protection, structural strength, and nuclear hardening of the interceptor structure.

The material design requirements or drivers result in materials with high specific strength and modulus to meet the minimum weight penalty. Table 2-1 summarizes the properties and characteristics of advanced composite materials for interceptor structural application. However, the material property requirements are not limited to standard mechanical characteristics such as longitudinal strength, transverse strength, shear strength, etc., but also include other required properties and characteristics such as coefficient of thermal expansion, specific heat, damping loss factor, laser hardness, etc., as shown in Table 2-1.

6

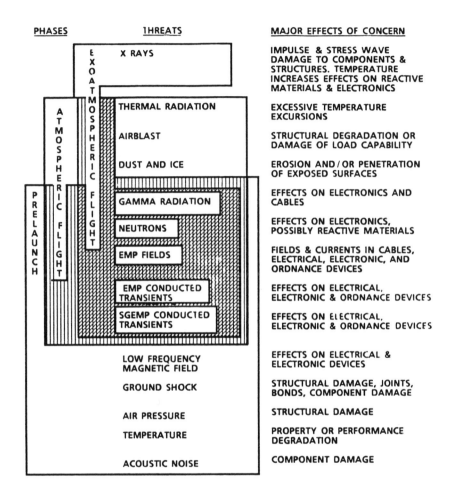

Figure 2-1. <u>Interceptor structural environmental threats.</u> (Reference 1)

TABLE 2-1. Material Selection Properties and Characteristics.

STATIC CHARACTERISTICS
   LONGITUDINAL STRENGTH
   TRANSVERSE STRENGTH
   SHEAR STRENGTH COMPRESSION
   YOUNG'S MODULUS
   POISSON'S RATIO
	DAMPING CHARACTERISTICS
	   LOSS FACTOR
	
	THERMAL PROPERTIES
	   COEFFICIENT OF THERMAL EXPANSION
	   HEAT TRANSFER COEFFICIENT
	   SPECIFIC HEAT

FATIGUE CHARACTERISTICS
   HIGH LOAD
   LOW LOAD / EXTENDED LIFE
   CONSTANT AMPLITUDE LOAD
   SPECTRUM LOAD
	MANUFACTURING METHODS
	   PRODUCIBILITY
	   PROCESSING CHARACTERISTICS
	   MINIMUM HANDLING THICKNESSES
	   JOINING TECHNIQUES
	   NDI METHODOLOGY
	   QUALITY ASSURANCE

FRACTURE CHARACTERISTICS
   FRACTURE TOUGHNESS
   FLAW GROWTH CHARACTERISTICS
	HOSTILE ENVIRONMENTS
	   MOISTURE
	   TEMPERATURE
	   NUCLEAR HARDNESS
	   LASER HARDNESS
	   BEAM WEAPON HARDNESS

2.2 Selection Factors.   The second material selection criterion is the material selection factors.   The selection factors for an advanced composite material system are summarized in Table 2-2.   As an example, some critical selection factors include an available data base, material availability on demand, and low material cost.   For the available data base factor, it should be noted that some of the material data are specific to certain applications and perhaps not necessarily of interest to USASDC. However, a complete material data base will include the material design, analysis, processing, and mechanical properties.   At the present, an important factor for the material data base is the general lack of information provided for the samples being tested and reported.   The quality and properties of a material vary not only with processing conditions, but also with time and probably some undefined variables.

Another important selection factor is the composite material cost. Presently, high cost is a primary barrier to large scale use of advanced composite material systems.   It results from high cost and structural fabrication cost of raw reinforcement materials.   It is expected that significant cost reduction will occur in the material quality control inspection and manufacturing of composite hardware with increased production.   These cost reductions will occur primarily because of increased automation, decreased raw material cost, and decreased cost as a result of the learning curve.

TABLE 2-2.   USASDC Advanced Material Selection Factors.

- **AVAILABLE DATA BASE**
  - **DESIGN**
  - **ANALYSIS**
  - **PROCESSING**
  - **MECHANICAL PROPERTIES**
- **MATERIALS AVAILABLE ON DEMAND**
- **LOW MATERIAL COSTS**
- **EASY TO MAKE**
- **RELIABLE**
- **EASY TO INSPECT**
- **HIGHER STRENGTH / DENSITY**
- **HIGHER STIFFNESS / DENSITY**

3. Applicable Materials

3.1 <u>Advanced Composite Components</u>. The major driving force for using advanced composite materials in interceptor structures is the superior mechanical properties of the composites. Composite materials generally consist of a bulk material called the matrix and a filler or reinforcement material of some type, such as fibers, whiskers, particulates, or fabrics. The composite materials are usually divided into three broad groups identified by their matrix materials: metal, polymer, or ceramic. With composite materials it is possible to tailor the properties of a component to meet the needs of a specific design by appropriate selection of matrix materials and the reinforcement agents. The composite concepts involve reinforcing matrices with a variety types of reinforcement materials are shown in Figure 3-1.

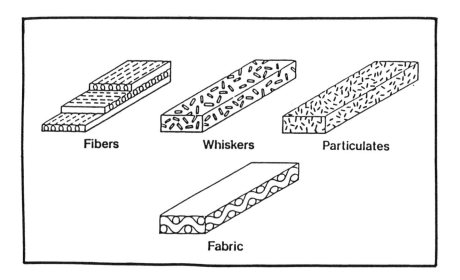

Figure 3-1. <u>Composite material approaches.</u>

The reinforcement materials consist of high strength materials in continuous fibers, whiskers, particulates or fabric form. These reinforcement materials usually carry the major stresses and loads, while the matrix material holds them together, enabling the stresses and loads to be transferred to the reinforcement materials. This is the case for high strength, filament-wound composite motor cases. The ability to tailor the properties is expanded by being able to select different reinforcements and matrices as shown in Tables 3-1 and 3-2.

TABLE 3-1.   Variety of Different Types of Reinforcements for Composite Materials.

REINFORCING AGENTS

CONTINUOUS FIBERS
- BORON (B)
- GRAPHITE (C)
- ALUMINA (Al_2O_3)
- SILICON CARBIDE (SiC)
- BORON CARBIDE (B_4C)
- BORON NITRIDE (BN)
- SILICA (SiO_2)
- TITANIUM DIBORIDE (TiB_2)
- ALUMINA-BORIA-SILICA ("NEXTEL")

WHISKERS

OVER 100 MATERIALS PRODUCED

METAL REINFORCEMENTS
- IRON (Fe)
- NICKEL (Ni)
- COPPER (Cu)
- NICKEL ALUMINIDE ($NiAl_3$)
- ALUMINUM OXIDE-ALUMINA-SAPPHIRE (Al_2O_3)
- SILICON CARBIDE (SiC)
- GRAPHITE (C)
- SILICON NITRIDE (Si_3N_4)

PARTICULATES (including flakes)
- TUNGSTEN (W)
- MOLYBDENUM (Mo)
- CHROMIUM (Cr)
- SILICON CARBIDE (SiC)
- BORON CARBIDE (B_4C)
- TITANIUM CARBIDE (TiC)
- ALUMINUM DODECABORIDE (AlB_{12})
- TUNGSTEN CARBIDE (WC)
- CHROMIUM CARBIDE (Cr_3C_2)
- SILICA (SiO_3)
- ALUMINA (Al_2O_3)
- MOLYBDENUM DISILICIDE ($MoSi_2$)

METAL WIRES
- TUNGSTEN (W)
- TITANIUM (Ti)
- MOLYBDENUM (Mo)
- BERYLLIUM (Be)
- STAINLESS STEEL
- NIOBIUM-TIN (NbSn) - SUPERCONDUCTOR
- NIOBIUM-TITANIUM (NbTi) - SUPERCONDUCTOR

TABLE 3-2.   Variety of Different Types of Matrix Materials
for Composite Materials.

MATRIX MATERIALS

METALLICS

ALUMINUM	SILVER
MAGNESIUM	ZINC
TITANIUM	BRONZE
COPPER	COBALT
NICKEL	IRON
LEAD	ALL ALLOYS OF ABOVE

PLASTICS	CERAMICS
EPOXIES	ALUMINUM OXIDE
POLYIMIDES	PORCELAIN
POLYSULFONES	PLASTER
POLYSTYRENES	CARBON
DIALLYL PHTHALATE	SILICON NITRIDE
PHENOLICS	
ARAMIDS	
POLYESTERS	
POLYCARBONATE	

3.2 Candidate Materials. This document provides a review of some of the most prominent metal matrix and polymer matrix composite materials. The material systems that have been chosen for this study are being seriously considered for engineering structural application to USASDC advanced material systems. As shown in Figure 1-1, graphite, boron, Kevlar, silicon carbide, and fiberglass are the principal reinforcement materials considered. Aluminum, magnesium, and titanium are the most important metal matrices. Epoxy, phenolic, and polyimide are the most important polymer matrices. Figures 3-2 and 3-3 show the variations of the specific strength (strength/density) and specific stiffness (modulus/density) properties with respect to temperature for some of the most prominent metal matrix and polymer matrix composite materials.

As seen in Figures 3-2 and 3-3, the polymer matrix composites such as graphite-epoxy and graphite-polyimide provide strength and stiffness properties for low temperature applications only. However, with proper matrix and reinforcement selection and design, it is possible that the polymer composites may provide significant advancement for advanced interceptor applications.

The whisker reinforcement system involving metal matrices provides significantly better strength and stiffness properties at higher temperatures when compared with polymer matrix composites. They also provide more of the desired properties such as electrical, thermal conductivity, and radiation resistance that are available from conventionally metallic structures. The metal matrix composites with continuous fiber reinforcements have potentially greater application than whisker reinforcement systems. However, metal matrix composite development is at approximately the same state as polymer matrices were about 15 years ago. Therefore, there is still much research and development required before these metal matrix composites will be available for large quantity use.

Figure 3-4 shows the relationship between the interceptor components, their structural requirements, and the potential application for candidate advanced composite materials. The interceptor structural components consist of the shroud, forecone, aftbody, bulkheads, heat shield, etc. Each of these structural components has its key design requirements such as high body strength and stiffness, high body bending frequency, minimum weight, hardness to nuclear and DEW, etc. As seen in Figure 3-4, the potential application for advanced structural materials in USASDC interceptor systems is found in numerous locations along the interceptor structure.

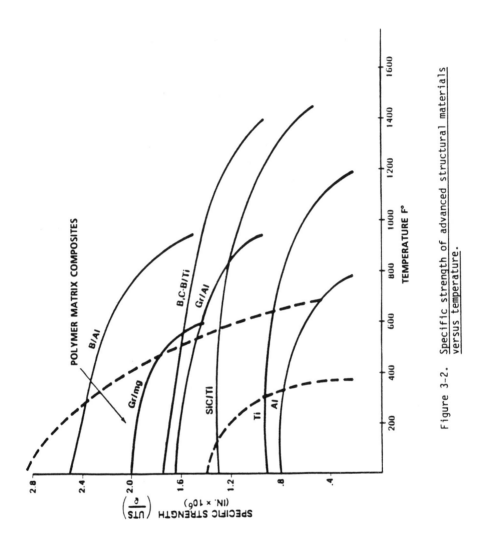

Figure 3-2. Specific strength of advanced structural materials versus temperature.

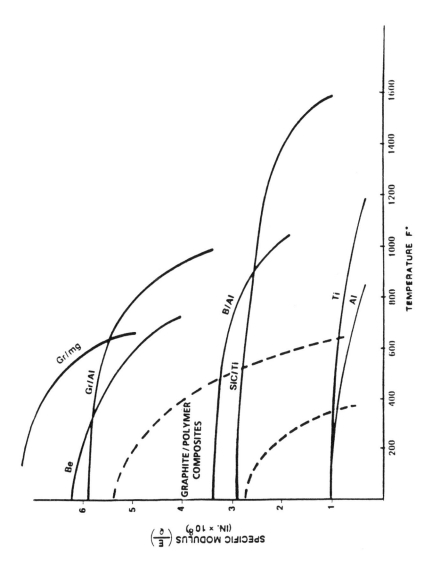

Figure 3-3. Specific stiffness of advanced structural material versus temperature.

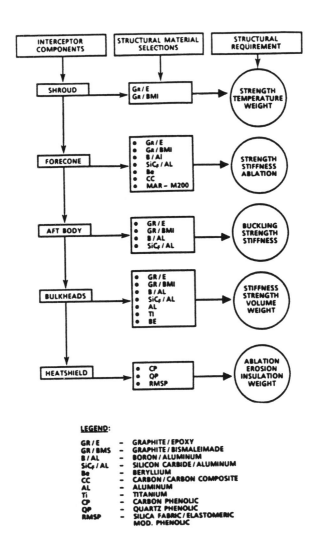

Figure 3-4.   Interceptor structural requirement and material
selections.

3.3 <u>Current Assessment for MMC Materials</u>.   The current technology assessment for metal matrix composites (MMC) materials is shown in Table 3-3.   The MMC development is at approximately the same stage as polymer matrices were about 15 years ago.   However, MMC materials using whisker reinforcements provide significantly better specific stiffness and higher specific strength at higher elevated temperature than polymer matrix composites.   At present, the MMC materials involve expensive and complex manufacturing methods.   In general, high cost is one of the primary barriers to large scale use of composite materials.

TABLE 3-3.   Current Assessment for MMC Materials.

TECHNOLOGY IN INFANCY – STILL EVOLVING.

GREAT TECHNICAL POTENTIAL.

COST IS THE KEY.

PLASTICITY EFFECTS NOT WELL DEFINED.

WHISKER AND PARTICULATE SYSTEMS LOOK GOOD.
    SIGNIFICANTLY BETTER STIFFNESS / DENSITY
    BETTER ELEVATED TEMPERATURE PROPERTIES
    ADAPTABLE TO CONVENTIONAL METAL FABRICATION METHODS
    LOW COST POTENTIAL

FUTURE OF CONTINUOUS FIBER SYSTEMS LESS CLEAR.
    EXPENSIVE AND COMPLEX FABRICATION METHODS
    HIGH TEMPERATURE RESIN SYSTEMS STRONG COMPETITION FOR ALUMINUM
    MATRIX COMPOSITES

CONCENTRATION OF FEW "HIGH PAYOFF" EXISTING SYSTEMS SHOULD BE PREFERRED OVER FRAGMENTED EFFORTS TO DEVELOP ENTIRELY NEW, UNPROVED SYSTEMS.

MMC ARE AT ABOUT THE SAME STAGE OF DEVELOPMENT THAT POLYMER MATRIX COMPOSITES WERE 15 YEARS AGO (BORON / ALUMINUM IS AN EXCEPTION).

The current MMC problems and disadvantages are associated with an immature technology which tends to make the risk in using MMC systems very high and, thus, limit their application in near-term systems. The advanced composite materials are still in the new technology stage and require significant research and development. Figure 3-5 presents the Department of Defense (DoD) technology base funding for metal matrix composites, polymer matrix composite (graphite/epoxy), and ceramic matrix composite (carbon-carbon) from 1970 to 1982. In general, the advanced composite materials are still in the early stage and require significant government funding to obtain the near term state-of-the-art advances necessary to meet the needs of USASDC advanced interceptors.

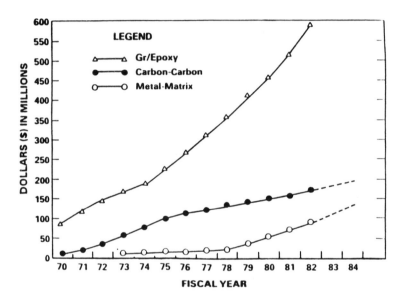

Figure 3-5.   DoD technology base funding for graphite/epoxy, carbon/carbon, and metal/matrix composites. (Reference 2)

4. Material Cost Projections

4.1 <u>Quantitative Costs</u>. Examination of the cost of using composite materials involves consideration of several factors. These factors include the cost of raw materials, the cost of processing the materials into composite preforms, and the cost of fabricating composite structures. Table 4-1 lists the approximate composite material costs derived using the volume-weighted averages technique (Reference 3). In general, the raw materials for composite systems are quite expensive when compared with monolithic structural materials. Therefore, the average composite material costs seem to decrease as the reinforcement material costs decrease. For example, the cost of a graphite-aluminum composite is mainly driven by the cost of the graphite fibers. However, it is expected that the reinforcement material costs will decline significantly because of increased fiber production rates which result from improvements in fabrication technology and from a learning curve phenomenon.

TABLE 4-1. Approximate Cost of Epoxy and Aluminum Composites.

REINFORCEMENT	MATRIX	FIBER VOLUME FRACTION (%)	DENSITY (LB / IN^3)	COST ($ / LB)
BERYLLIUM	EPOXY	48.4	0.062	2595.5
VHM FIBER	EPOXY	16.5	0.060	390.2
BORON	EPOXY	33.5	0.067	108.6
GRAPHITE	EPOXY	36.0	0.060	8.5
SILICON CARBIDE	EPOXY	51.5	0.088	4.6
BERYLLIUM	ALUMINUM	32.3	0.088	1215.5
VHM FIBER	ALUMINUM	8.3	0.096	126.0
BORON	ALUMINUM	25.6	0.095	61.2
GRAPHITE	ALUMINUM	22.2	0.091	6.5
SILICON CARBIDE	ALUMINUM	35.1	0.105	4.7

6027-6

19

Figure 4-1 shows the price per pound of graphite and boron fibers as a function of time. Graphite fiber has dropped its price significantly over the past 20 years. As a result, the cost of graphite-epoxy and graphite-aluminum composites could be obtained at $8.50 per pound and $6.50 per pound in 1985 dollar value, respectively. These values are taken from Table 4-1. From the result of Figure 4-1, boron fiber cost is still higher than graphite fiber, and thus boron filament cost is the significant factor associated with the mass production of boron composite materials such as boron-epoxy or boron-aluminum. From Table 4-1, the potential low cost of silicon carbide reinforced aluminum composite is one of the most attractive features of these advanced composite materials.

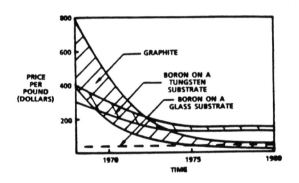

Figure 4-1.   Graphite and boron fibers cost projections.   (Reference 4)

Table 4-2 shows the advanced composite materials in terms of their cost-value relationship. The significant aspect of the cost-value relationship is expressed in terms of the material specific strength (strength per density) and material specific stiffness (modulus per density) per unit material cost. As seen from this table, silicon carbide reinforced epoxy gives the high specific strength pay-off, since its specific strength value is about $2,012 \times 10^3$ inch per dollar, and its specific stiffness is about 656×10^6 inch per dollar. Although silicon carbide-epoxy provides the highest values of specific properties per cost, the composite can be used for low temperature (less than 350 °F) application only. This service limitation is caused by the fact that epoxy is a polymer material. On the other hand, silicon-carbide reinforced aluminum can provide moderately high specific values at a much higher service temperature.

TABLE 4-2. Composite Material Value-Cost Relationships.

REINFORCEMENT MATERIAL	MATRIX MATERIAL	COST ($/LB)	UTSL (KSi)	STRENGTH/COST (KSi/$)	MODULUS/COST (MSi/$)	SPECIFIC STRENGTH COST (x 10³ IN/$)	SPECIFIC STIFFNESS COST (x 10⁶ IN/$)
BERYLLIUM	EPOXY	2595.5	74.6	0.46	0.12	7.40	1.98
VHM FIBER	EPOXY	390.2	55.6	2.37	0.85	39.51	14.19
BORON	EPOXY	108.6	145.5	19.78	2.72	292.20	40.16
GRAPHITE	EPOXY	8.5	61.3	120.30	39.30	2,012.1	656.55
SILICON CARBIDE	EPOXY	4.6	41.7	104.1	49.86	1,184.4	567.30
BERYLLIUM	ALUMINUM	1215.5	58.6	0.54	0.18	6.14	2.10
VHM FIBER	ALUMINUM	126.0	38.1	3.13	1.64	32.36	16.98
BORON	ALUMINUM	61.2	122.9	20.93	3.40	218.3	35.52
GRAPHITE	ALUMINUM	6.5	46.6	79.30	34.01	872.02	373.80
SILICON CARBIDE	ALUMINUM	4.7	65.0	129.7	39.90	1,234.9	379.90

UTSL = ULTIMATE TENSILE STRENGTH (LONGITUDINAL)
SPECIFIC STRENGTH = STRENGTH/DENSITY
SPECIFIC MODULUS (STIFFNESS) = MODULUS/DENSITY
VHM = VERY HIGH MODULUS FIBER

4.2 <u>Structural Projected Costs</u>.   The costs of sophisticated
structures such as those found in the missile interstages and payload
structures of an advanced interceptor were estimated based upon satellite
structure cost analysis (Reference 5).   Table 4-3 shows the costs for the
years 1980 through 2000.   For advanced metal matrix composites these
estimations are based on the assumption that MMC will grow to maturity at
about the same rate as did polymer matrix composites (boron/epoxy and
graphite/epoxy).   The projections were made in early 1983, and they include
an estimation for inflation, which may be conservative based upon the 1985
rates.   As a result of inflation, Table 4-3 shows that aluminum structural
costs would increase from $10 per pound to $15 per pound from the 1980's to
the 1990's, respectively.

In estimating the <u>structural</u> cost for a meteorological satellite using
advanced composite material (graphite/magnesium), it is found that the
total <u>material</u> cost is still less than one percent of the total structural
cost (Reference 5).   This result also can be found when comparing, as an
example, graphite/aluminum <u>structural</u> cost ($4000 per pound) in Table 4-3
and graphite/aluminum <u>material</u> cost ($6.50 per pound) in Table 4-1.   Both
cost values are expressed in 1985 dollar value.   <u>This suggests that for the</u>
<u>USASDC interceptor materials development goals, the current and projected</u>
<u>material costs should be considered secondary to the technical gains that</u>
<u>might be achieved in an advanced interceptor</u>.

TABLE 4-3.   Cost Projections for the Candidate Material
Technologies.   (Reference 5)

MATERIAL	1980's COST ($ / LB)	EQUIVALENT STRUCTURAL WEIGHT (LB)	SATELLITE STRUCTURE COST ($)
Al	$10 / LB	250	$2.5K
GR / EP	$500 / LB	58	$29K
GR / Al	$4000 / LB	50	$200K
GR / MG	$6000 / LB	45	$270K

MATERIAL	1990's COST ($ / LB)	EQUIVALENT STRUCTURAL WEIGHT (LB)	SATELLITE STRUCTURE COST ($)
Al	$15 / LB	250	$3.75K
GR / EP	$400 / LB	41	$16K
GR / Al	$1200 / LB	34	$41K
GR / MG	$1800 / LB	29	$52K

5. Conclusion

5.1 <u>General</u>.

(a) The key design requirements for advanced interceptors include high body bending frequency, high body strength, stiffness, low body weight, and hardness to nuclear and DEW's. These design requirements and material selection factors are the material selection criteria for USASDC advanced composite material systems.

(b) Graphite, boron, Kevlar, silicon carbide, and fiberglass are the principal reinforcement materials. Aluminum, magnesium, and titanium are the most important metal matrices. Epoxy, phenolic, and polyimide are the most important polymer matrices.

(c) Because of low temperature and low cost fabrication methods, polymer matrix composite development has maintained a distance ahead of metal matrix composite. Initial skepticism of polymer matrix composite has faded, and it is now a question of where, rather than whether, to use polymer matrix composites for advanced interceptor structural application.

(d) The MMC development is at approximately the same stage as polymer matrices were about 15 years ago. The MMC materials are still in the early stage and require significant government funding to obtain the near-term state-of-the-art advances necessary to meet the needs of USASDC advanced interceptors.

5.2 <u>Barriers to Large Scale Use of Composites</u>.

(a) High cost is a primary barrier. In general, the reinforcement materials cost for advanced composite systems are quite expensive and, therefore, the average composite material costs seem to decrease as the reinforcement material costs decrease.

23

(b) Unavailability of a large material data base is another barrier
     to large scale use of composites. It should be noted that some
     of the material data are specific to certain applications and
     perhaps not necessarily of interest to USASDC material systems.
     However, a complete material data base will include the material
     design, analysis, processing, and mechanical properties. It is
     also recognized that some processing information is proprietary
     to the supplier. This problem could be the cause for lack of
     adequate quality control methods for raw materials and composite
     fabricated structures.

6. Recommendation

In general, the MMC materials are still in the early stage of development and require significant government funding to obtain near-term state-of-the-art advances necessary to meet the needs of USASDC advanced interceptors. The high cost of advanced composite materials is a primary barrier to large scale use of composite structures; continuing attention should be paid to decreasing the costs of production of the raw materials, in this case, the cost of the reinforcement materials and fabrication.

Laboratory projects on advanced metal matrix composites systems should be initiated on a priority basis. The following advanced composite materials are recommended for strong research funding, and the work should be accelerated since the long-term payoff for these materials can be quite large. For metal matrix composites they are: graphite/aluminum, silicon carbide/aluminum, graphite/magnesium and boron/titanium. For polymer matrix composites, they are: graphite/epoxy, boron/epoxy, Kevlar/epoxy, graphite/polyimide, fiberglass/phenolic, and graphite/phenolic.

The unavailability of a large composite material data base is a barrier to large scale use of composite structures. It is recommended that a complete material data base which includes the material standardized design allowables (as in Mil-HdbK-5 and 17), material analysis, processing, and mechanical properties are needed to facilitate the advanced material selection for USASDC material systems.

25

Appendix: Material Property Data Summary

The superior mechanical properties of composite materials is one of the major driving forces for their use. An important characteristic of composite materials is that by appropriate selection of matrix materials and reinforcement fibers, it is possible to tailor the properties of a component to meet the needs of a specific design. Because of the low temperature and low cost fabrication method, polymer matrix composite development has maintained a lead on metal matrix composites. In essence, polymer matrix composites result in materials that have higher specific stiffness, specific strength, permit more flexible design, and are more easily repaired. However, polymer matrix composites can only be applied for low service temperature (less than 600 °F).

Metal matrix composites are superior under compressive buckling loads because of the higher modulus of the metal matrices. Metal matrix composites are more erosion resistant and have higher service temperatures. Their good thermal conductivity, high electrical conductivity, and low thermal expansion are particularly attractive for advanced interceptor structural applications. However, metal matrix composite technology is in the early stage of development and the fabrication costs are considerably higher than polymer matrix composite.

Table A-1 and Figure A-1 show the representative properties of metal matrix composites in comparison with properties of polymer matrix composites (epoxy). Other typical properties of metal matrix and polymer matrix composites can be found from Figure A-2 to Figure A-7. Volume II gives more detailed information on the advanced composite mechanical, thermal, and physical properties.

TABLE A-1.   Representative Properties of Metal Matrix Composites
(Reference 6)

Matrix	Reinforcement	Reinforcement (Volume Percent)	Modulus (10^6 psi)		Tensile Strength (10^3 psi)	
			Longitudinal	Transverse	Longitudinal	Transverse
Aluminum	None	0	10	10	40-70	40 70
Epoxy	High-strength graphite fibers	60	21	1.5	180	6
Aluminum	Alumina fibers	50	29	22	150	25
Aluminum	Boron fibers	50	29	18	190	15
Aluminum	Ultrahigh modulus graphite fibers	45	50	5	90	5
Aluminum	Silicon carbide particles	40	21	21	80	80
Titanium	Silicon carbide monofilament fibers	35	31	24	250	60

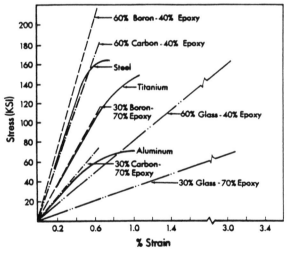

Figure A-1.   Comparison of epoxy materials with steel, titanium, and aluminum.   (Reference 7)

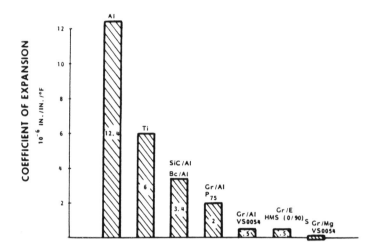

Figure A-2.   Comparison of thermal coefficient of expansion for composite materials.   (Reference 8)

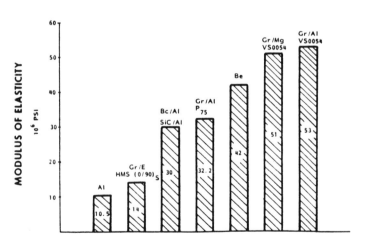

Figure A-3.   Comparison of modulus of elasticity for composite material.   (Reference 9)

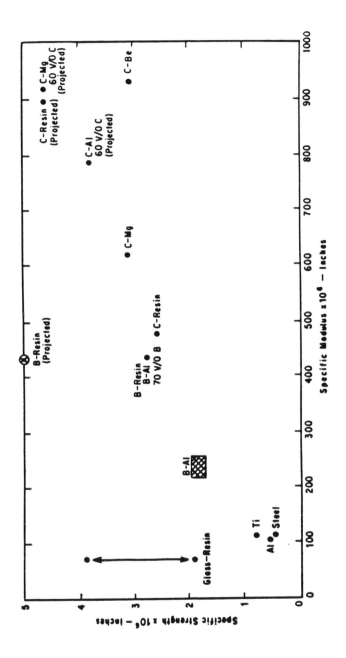

Figure A-4.   Comparison of specific properties of metal and polymer composites. (Reference 10)

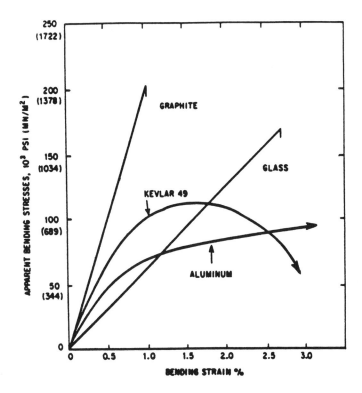

Figure A-5. <u>Unidirectional composite bending stress versus strain curves for various reinforcement materials in epoxy resin.</u> (Reference 11)

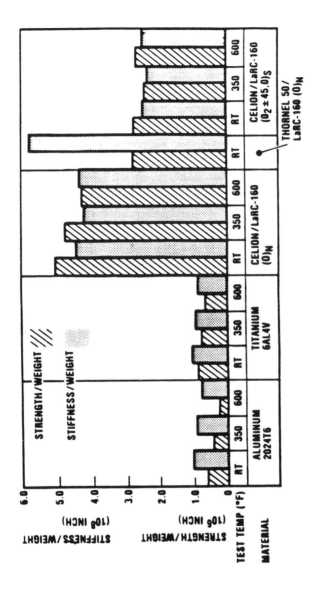

Figure A-6.   Graphite/polyimide composite system capable of
600 F° service temperature.   (Reference 12)

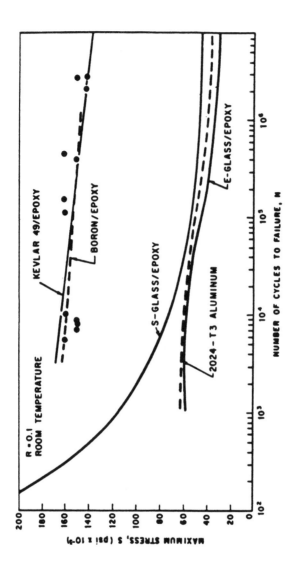

Figure A-7. Tension fatigue behavior of unidirectional Kevlar/epoxy composites. (Reference 13)

References

1. Garcia, D., Willis, J., Muha, T., "Organic Matrix Composite Application to Missile Structure," in <u>Fibrous Composites in Structural Design</u>, Plenum Press, New York, 1980.

2. Persh, J., <u>The Department of Defense Metal Matrix Composites Technology Thrust - Five Years Later</u>, paper presented at the Fifth Metal Matrix Composites Technology Conference, Silver Spring, MD, May 1983.

3. Rubin, L., <u>Performance and Cost Implication of Metal Matrix Composite Structures in Satellite Systems</u>, paper presented at the 25th Structure, Structural Dynamics and Materials Conference, Palm Springs, CA, May 1984.

4. Jones, R. M., <u>Mechanics of Composite Materials</u>, McGraw-Hill Book Company, New York, 1975.

5. Rubin, p. 9.

6. Zweben, C., "Metal Matrix Composites Overview," MMCIAC Publications, No. 253, February, 1985.

7. McCullough, R., <u>Concepts of Fiber-Resin Composites</u>, Marcel Dekker, New York, 1971.

8. Armstrong, H., <u>Metal Matrix Composites for Space Application</u>, paper presented at the Fifth Metal Matrix Composites Technology Conference, Silver Spring, MD, May 1983.

9. Armstrong, H., p. 23-5

10. Watts, A. A., <u>Commercial Opportunities for Advanced Composites</u>, STP-704, ASTM special pub., 1980.

11. Watts, A. A., p. 60.

12. Jones, J., "Cellion/LaRC-160 Graphite/Polyimide Composite Processing Techniques and Properties," <u>SAMPE Journal</u>, April 1983.

13. Watts, A. A., p. 63.

Part II

Materials Properties Review

The information in Part II is from *Advanced Materials Research Status and Requirements. Volume II—Appendix: Material Properties Data Review,* prepared by Jonathan A. Lee and Donald L. Mykkanen of BDM Corporation for the U.S. Army Strategic Defense Command, March 1986.

1. Introduction

1.1 <u>Purpose and Scope of Document</u>. This document is Volume II of a two-volume report describing the status and requirements of the advanced composite materials for the U.S. Army Strategic Command (USASDC). It has been prepared to provide material property information for the selection of advanced composite material systems. The information contained in this volume is the result of a thorough search of the Defense Technical Information Center (DTIC) literature, contractor reports, the Metal Matrix Composites Information Analysis Center (MMCIAC), and open literature. This search covers the years 1975 to mid-1984. The objective of Volume II is to present data on the mechanical, thermal, and physical properties of general interest advanced metal matrix and plastic (polymer) systems. Because advanced composite materials are in a state of evolution in terms of property improvements, it is not possible to provide final property values in the same sense as those now available for conventional metal alloys. Thus, this volume is intended to inform the reader in general terms rather than to serve as a standard sourcebook for the advanced composite systems.

1.2 <u>Advanced Composite Material Overview</u>. Composite materials generally consist of a bulk material, called the matrix, and a filler or reinforcement material such as fibers, whiskers, or metal wires. The composite materials are usually divided into three broad groups identified by their matrix materials: metal, plastic (polymer), or ceramic. The reinforcement materials or agents consist of high-strength materials in continuous fibers, whiskers, particulates, or fabric form. These reinforcement materials usually carry the major stresses and loads, while the matrix material holds the reinforcement materials together. Figure 1.2-1 shows schematically how a composite material is selected and created. Because of the superior mechanical and physical properties of composite materials, they are attractive for advanced structural and nonstructural applications.

Metal matrix composite (MMC) technology began in the late 1950's and is still in the early stages of development. Many different combinations

of matrices and reinforcements have been investigated and more seem immi-
nent.   There are four major categories of composite reinforcements: continuous fibers, whiskers, particulates, and wires or fabric.   Table 1.2-1
presents the variety and complexity of different types of reinforcement
materials.

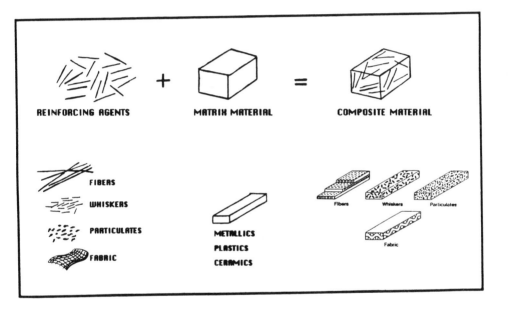

Figure 1.2-1   Composite Material Overview

TABLE 1.2-1.   Variety of and Complexity of Different
              Types of Reinforcement Materials for
              Composite Materials

REINFORCING AGENTS

CONTINUOUS FIBERS
  BORON (B)
  GRAPHITE (C)
  ALUMINA (Al_2O_3)
  SILICON CARBIDE (SiC)
  BORON CARBIDE (B_4C)
  BORON NITRIDE (BN)
  SILICA (SiO_2)
  TITANIUM DIBORIDE (TiB_2)
  ALUMINA-BORIA-SILICA ("NEXTEL")

WHISKERS

  OVER 100 MATERIALS PRODUCED

METAL REINFORCEMENTS
  IRON (Fe)
  NICKEL (Ni)
  COPPER (Cu)
  NICKEL ALUMINIDE ($NiAl_3$)
  ALUMINUM OXIDE-ALUMINA-
     SAPPHIRE (Al_2O_3)
  SILICON CARBIDE (SiC)
  GRAPHITE (C)
  SILICON NITRIDE (Si_3N_4)

PARTICULATES (including flakes)
  TUNGSTEN (W)
  MOLYBDENUM (Mo)
  CHROMIUM (Cr)
  SILICON CARBIDE (SiC)
  BORON CARBIDE (B_4C)
  TITANIUM CARBIDE (TiC)
  ALUMINUM DODECABORIDE (AlB_{12})
  TUNGSTEN CARBIDE (WC)
  CHROMIUM CARBIDE (Cr_3C_2)
  SILICA (SiO_3)
  ALUMINA (Al_2O_3)
  MOLYBDENUM DISILICIDE ($MoSi_2$)

METAL WIRES

  TUNGSTEN (W)
  TITANIUM (Ti)
  MOLYBDENUM (Mo)
  BERYLLIUM (Be)
  STAINLESS STEEL
  NIOBIUM-TIN (NbSn) -
     SUPERCONDUCTOR
  NIOBIUM-TITANIUM (NbTi) -
     SUPERCONDUCTOR

Common continuous fibers include boron, graphite, alumina, and silicon carbide. Boron fibers are manufactured by chemical vapor deposition (CVD) onto a tungsten or carbon core. In many applications, a coating of silicon carbide is placed on the boron fibers to avoid reactions with the matrix. Silicon carbide fibers are made in a CVC process similar to that of boron. Continuous alumina fibers have properties that vary significantly from manufacturer to manufacturer. There are two types of graphite fibers, one made from pitch, the other from polyacrilonitrile (PAN). Graphite fibers are available in a wide variety of strengths and moduli.

Metal matrix composites have been tested with many metals, but the most important are aluminum, titanium, magnesium, and copper alloys. Aluminum matrix composites are reinforced with continuous fibers such as boron, silicon carbide, alumina, and graphite; discontinuous fibers such as alumina and alumina-silca; silicon carbide whiskers; or particulates of silicon carbide and boron carbide. Magnesium matrices are reinforced with graphite or alumina fibers, silicon carbide whiskers, and silicon carbide or boron carbide particulates. Titanium systems include reinforcements by coated boron fibers, silicon carbide filters, and titanium carbide particu- lates. Copper matrix composites may be reinforced with graphite or silicon carbide fibers; niobium-titanium or niobium-tin wires; or silicon carbide, boron carbide, or titanium carbide particulates.

Table 1.2-2 shows the variety of types of matrix materials for compos- ite materials. Several other composite systems have received lesser interest to date, and research in these systems is proceeding at a slower pace.

This report will review metal matrix and polymer matrix composite sys- tems. The emphasis will be on composite systems that have been tested for structural applications, rather than on composites tested solely for behavior patterns.

TABLE 1.2-2. Variety of Different Types of Matrix Materials
              for Composite Materials

MATRIX MATERIALS

METALLICS

ALUMINUM	SILVER
MAGNESIUM	ZINC
TITANIUM	BRONZE
COPPER	COBALT
NICKEL	IRON
LEAD	ALL ALLOYS OF ABOVE

PLASTICS	CERAMICS
EPOXIES	ALUMINUM OXIDE
POLYIMIDES	PORCELAIN
POLYSULFONES	PLASTER
POLYSTYRENES	CARBON
DIALLYL PHTHALATE	SILICON NITRIDE
PHENOLICS	
ARAMIDS	
POLYESTERS	
POLYCARBONATE	

Figure 1.2-2 shows the advanced material selection recommended for the USASDC material program study.   Graphite, boron, Kevlar, silicon carbide, and fiberglass are the principal reinforcement materials considered in this document.   Although not truly an advanced reinforcement, fiberglass is included because of its broad, extensive use in military and commercial systems and products.   For metal matrix composites, aluminum, magnesium, and titanium are selected to be the most important metal matrices for USASDC material systems. For polymer matrix composites, epoxy, phenolic, and polyimide are selected to be the most important polymer matrices.

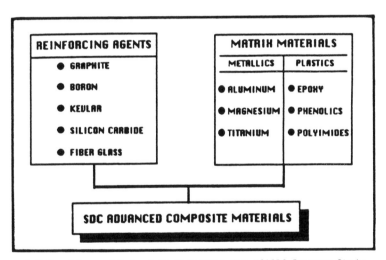

Figure 1.2-2.   Advanced Materials Selection for USASDC Program Study

1.3 <u>Metal Versus Plastic Composite Materials</u>. Metal matrix composites (MMCs) have advantages over monolithic metals which include better fatigue resistance, better wear resistance, better elevated temperature properties, higher strength-to-density ratios, higher stiffness-to-density ratios, and lower coefficients of thermal expansion. MMCs also have advantages when compared to polymer matrix composites, including higher functional temperatures, higher transverse strength and stiffness, no moisture absorption, better conductiveness, and better radiation resistance. MMCs also have disadvantages including high costs newer technologies, complex fabrication methods, and limited service experience.

Polymer matrix composites are highly anisotropic. Strength and stiffness are high parallel to the fibers, but low perpendicular to the fibers. PMCs have stress-strain curves that are generally linear to failure. Polymer matrix materials result in composites that have higher specific tensile strength and stiffness properties. Plastic composites are more advanced in the fabrication technology and are lower in raw material cost and fabrication cost.

Both metal and polymer matrix properties are affected by reinforcement properties, form, and arrangement; reinforcement volume; matrix properties; and reinforcement - matrix interface properties. They may also be affected by residual stress and degradation because of high temperatures and mechanical damage.

2. Reinforcement Materials and Fabrication Methods

2.1 <u>Introduction</u>. Advanced composite materials are composed of metal, plastic, or ceramic matrix composite materials reinforced with high-strength, high-modulus fibers. This reinforcement may be continuous in the form of fibers or filaments, or discontinuous in the form of chopped fibers or whiskers. The main focus of this discussion will be on continuous reinforcement fiber materials. Advanced reinforcement materials include high-strength, high-modulus graphite, boron, glass, silicon carbide, and aramid fibers. Fiberglass is included, although it is not a true advanced reinforcement, because of its widespread use in military and commercial applications.

Fiber reinforcement is available in a wide range of sizes. Diameters vary from a few microns to several mils, lengths from long continuous fibers to short whiskers; the composite material may contain only a few fibers, or its fiber content may be as much as 80 percent. Figure 2.1-1 shows a comparison of various fiber dimensions. This wide variety in fiber reinforcement makes many different handling techniques necessary. Other unique properties may be found in Tables 2.1-1 and 2.1-2 and in Figures 2.1-2 and 2.1-3.

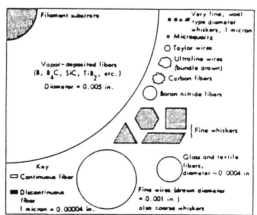

Figure 2.1-1. <u>Relative Cross-Sectional Areas and Shapes of Several Fibrous Reinforcements</u> (Reference 2-1)

43

TABLE 2.1-1. Properties of Fibers and Conventional Bulk Materials
              (Reference 2-2)

Material	Tensile Modulus (E) (GN/m²)	Tensile Strength (σ_u) (GN/m²)	Density (ρ) (g/cm³)	Specific Modulus (E/ρ)	Specific Strength (σ_u/ρ)
Fibers					
E-Glass	72.4	3.5[a]	2.54	28.5	1.38
S-Glass	85.5	4.6[a]	2.48	34.5	1.85
Graphite (high modulus)	390.0	2.1	1.90	205.0	1.1
Graphite (high tensile strength)	240.0	2.5	1.90	126.0	1.3
Boron	385.0	2.8	2.63	146.0	1.1
Silica	72.4	5.8	2.19	33.0	2.65
Tungsten	414.0	4.2	19.30	21.0	0.22
Beryllium	240.0	1.3	1.83	131.0	0.71
Kevlar-49 (aramid polymer)	130.0	2.8	1.50	87.0	1.87
Conventional materials					
Steel	210.0	0.34-2.1	7.8	26.9	0.043-0.27
Aluminum alloys	70.0	0.14-0.62	2.7	25.9	0.052-0.23
Glass	70.0	0.7-2.1	2.5	28.0	0.28-0.84
Tungsten	350.0	1.1-4.1	19.30	18.1	0.057-0.21
Beryllium	300.0	0.7	1.83	164.0	0.38

[a]Virgin strength values. Actual strength values prior to incorporation
into composite are approximately 2.1 GN/m².

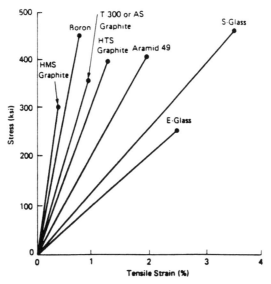

Figure 2.1-2. Stress-Strain Diagrams of Various Fibers (Reference 2-3)

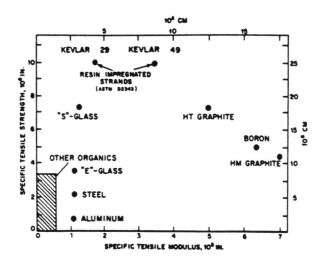

Figure 2.1-3. Specific Tensile Strength and Modulus of Reinforcing Fibers
(tensile strength or modulus divided by density)
(Reference 2-4)

TABLE 2.1-2.　Suppliers of Reinforcing Fiber and Preform Materials (Reference 2-5)

Fibers	Supplier, United States
Graphite	Hercules, Union Carbide, Celanese, Great Lakes Carbon, Stackpole Carbon
Boron/borsic	AVCO, Composite Materials Corporation
Silicon carbide filament	AVCO
Silicon carbide whiskers	EXXON
Fiberglass (E, S, quartz)	Owens Corning Fiberglass Company, PPG Industries
Aluminum oxide filament	TYCO, DuPont Company
Aramids (Kevlar) and other polymers	DuPont Company, 3M Company
Preforms and Prepregs	
Graphite	Fiber Materials, Inc., The 3M Company, U.S. Polymeric, NARMCO, Composite Materials Corporation, Ferro Corporation
Boron/borsic	Composite Materials Corporation
Fiberglass	Fiberite, U.S. Polymeric, NARMCO, Composite Materials Corporation, 3M Company
Graphite/aluminum	Fiber Materials, Inc.
Boron/borsic aluminum	DWA Composite Specialties, Inc., General Technologies Corporation

2.2　Properties of Reinforcement Materials.

2.2.1　Graphite Fibers. High-strength, high-modulus graphite fibers are manufactured by treating an organic base material fiber (precursor) with heat and tension, leaving a highly ordered carbon structure. The most commonly used precursors include rayon-base fibers, polyacrylonitride (PAN), and pitch. The following paragraphs present an introduction to the various processes involved with each precursor. Typical properties of the representative classes of graphite (carbon) fibers are listed in Table 2.2-1, and typical properties of specific commercially available carbon fiber materials are listed in Table 2.2-2. Other unique properties of carbon fiber materials can be found in Tables 2.2-3 and 2.2-4 and Figures 2.2-1 and 2.2-2.

2.2.1.1　Rayon-Base Fibers. Rayon carbon filaments are stretched in a series of steps at temperatures of around 2700 °C (4900 °F) while in an inert atmosphere. The tension, along with the high temperatures, causes the graphite layer planes to align with the fiber axis, giving them the high-strength, high-modulus characteristics.

2.2.1.2 Polyacrylonitrile (PAN) Carbon Fibers. PAN is a long chain linear polymer composed of a carbon backbone with attached carbonitrile groups. Fibers are made by first holding a stretched polymer in an oxidizing environment at 205 to 240 °C (400 to 460 °F) for 24 hours. This cyclization process produces a "ladder" type structure. The fiber is then heated in an inert atmosphere to temperatures ranging from 1400 to 3000 °C (2550 to 5400 °F), depending on desired characteristics.

2.2.1.3 Pitch Carbon Fibers. Coal tar pitch is heated for up to 40 hours at approximately 450 °C (800 °F). This forms a viscous liquid with a high degree of molecular order known as mesophase. The mesophase is then spun through a small orifice, aligning the mesophase molecules along the fiber axis. The fibers are thermoset at relatively low temperatures and then heat treated at 1700 to 3000 °C (3100 to 5400 °F).

TABLE 2.2-1.   Comparative Graphite Fiber Properties
(Reference 2-6)

Property	Pitch	Rayon	PAN
Tensile strength, 10^3 psi	225	300 to 400	450 to 360
Tensile modulus, 10^6 psi	55	60 to 80	30 to 50
Short beam shear, 10^3 psi			
untreated	6	4	10 to 4
treated	10	8	18 to 8
Specific gravity	2.0	1.7	1.8
Elongation, %	1	. . .	1.2 to 0.6
Fiber diameter, μm	. . .	6.5	7.5

TABLE 2.2-2.   Commercial Graphite Fiber Properties
(Reference 2-7)

Fiber	Precursor	Average Strength, 10^3 psi	Modulus, 10^6 psi	Density, g/cm^3
Thornel 50	rayon	320 ± 20	57 ± 3	1.67
Thornel 75	rayon	385 ± 17	76 ± 3	1.82
Modmor 1	PAN	340	59	1.85
Celion 70	PAN	300	88	1.96
Courtaulds HMS	PAN	250 − 350	50 − 60	1.94
Fortafil 6T	PAN	420 ± 20	59 ± 3	1.91
Fortafil 5T	PAN	400 ± 20	48 ± 3	1.80
Courtaulds HT	PAN	325 to 375	35 to 40	1.86
Thornel 300	PAN	325	34	1.70
Hercules Type A	PAN	400	28 to 32	1.77
UCC Type P	pitch	160 to 200	50	...

TABLE 2.2-3.   Thermal Conductivities of Different Carbon Fibers
(Reference 2-8)

Carbon fiber	Thermal conductivity (ω cm^{-1} k^{-1})
HM (PAN)	0.594
Type I (PAN)	1.02
Type II (PAN)	0.22
Thornel 50 (RAYON)	0.90 (1800 K)
Thornel 50 (RAYON)	0.60 (room temp.)
Morganite	0.20

TABLE 2.2-4.   Comparison of Mechanical Properties of Carbon Fibers with Other Fibrous Reinforcing Materials (Reference 2-9)

Fiber	Specific gravity (SG)	Ultimate tensile strength (UTS) ($GN\ m^2$)	Relative specific strength (UTS/SG)	Young's modulus (E) ($GN\ m^2$)	Relative Young's modulus (E/SG)
Carbon fiber Type I	1.90	2.6	1.37	340	179
Carbon fiber Type II	1.77	2.9	1.64	240	136
Carbon Fiber Type III	1.76	2.6	1.48	190	108
E-glass	2.59	3.44	1.38	72	29
S-glass	2.49	2.60	1.06	80	32
Kevlar 29	1.44	2.7	1.80	60	42
Kevlar 49	1.45	2.7	1.86	130	90
Boron (on tungsten)	2.50	3.75	1.50	400	160
Asbestos crocidolite	3.40	3.50	1.03	190	58
Asbestos chrysotite	2.50	3.11	1.24	160	64
Silicon carbide (on tungsten)	3.50	3.50	1.00	400	114

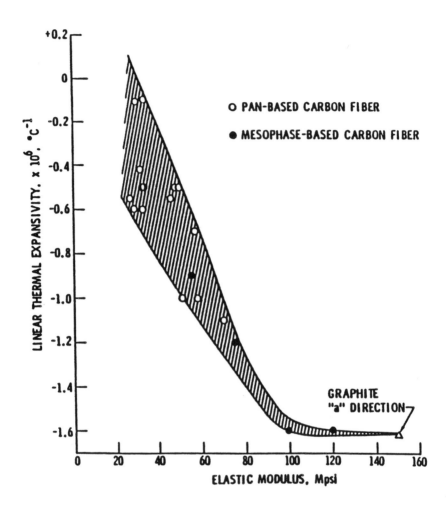

Figure 2.2-1.   Linear Thermal Expansivity of Carbon Fiber
(Reference 2-10)

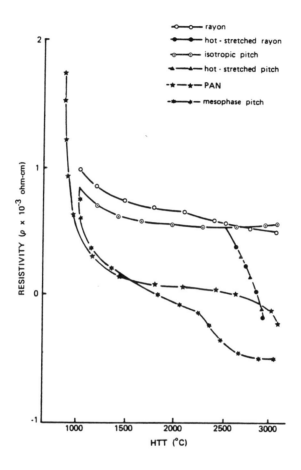

Figure 2.2-2.   Resistivity of Different Carbon Fibers as a Function of Heat Treatment Temperature   (Reference 2-11)

2.2.2 Boron Fibers. Boron fiber is made by deposits of chemical vapor onto a tungsten or carbon filament substrate. The filaments are heated to approximately 1260 °C (2300 °F) and then pulled through a boron trichloride/hydrogen environment. This deposits boron on the substrate at approximately 50mm per hour. Borsic filaments are made by depositing a layer of silicon dioxide onto the boron fiber in a similar process.

Table 2.2-5 shows fiber properties. Removing the tungsten core by continuous splitting and etching may produce a filament strength well above 6.895GPa (1,000,000 psi). Boron fibers have been produced for many years, and the accumulation of data is expected to further reduce costs and raise production volume. This has already been shown by the substitution of a carbon filament for tungsten. Other unique boron material properties can be found in Figures 2.2-3 through 2.2-5.

TABLE 2.2-5.   Room Temperature Mechanical Properties of Boron-on-Tungsten
                Filament (Reference 2-12)

Ultimate tensile, 10^3 psi	500+	500+	500+
Tensile modulus, 10^6 psi	58	58	58
Shear modulus, 10^6 psi	24	24	. . .
Coefficient of thermal expansion, in./in./°F $\times$ 10^6	2.7	2.7	2.7
Density, lb/in.^{3}	0.094	0.089	0.086
Length/pound, ft	70 000	37 800	19 000

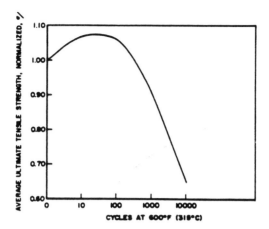

Figure 2.2-3.   Boron Filament Strength as a Function of
Cyclic Heat Treatment (Reference 2-13)

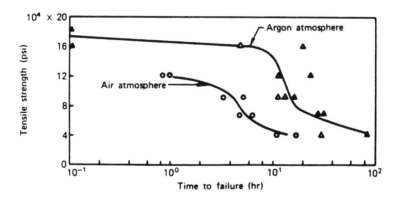

Figure 2.2-4.   Stress-Rupture Properties of Boron Fibers   (Reference 2-14)

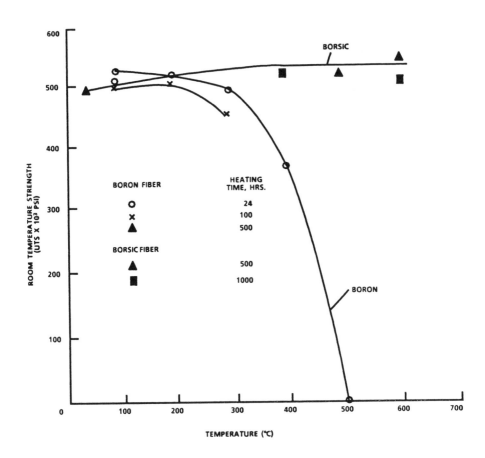

Figure 2.2-5.   Room Temperature Strength of Boron and Borsic Fibers
Heated in Air   (Reference 2-15)

2.2.3 Kevlar Fibers. Kevlar aramid fibers possess high strength and high modulus properties. Table 2.2-6 shows the properties of three Kevlar materials, and Figure 2.2-6 shows the effect of stress on the creep of Kevlar 49. The negative coefficient of thermal expansion must be considered when using Kevlar materials. Kevlar materials are available in yarns, rovings, staple, and fabrics. Table 2.2-7 shows the chemical resistance of Kevlar 49, the most advanced Kevlar material. Figures 2.2-7 and 2.2-8 show the heat and moisture effects on Kevlar 49. The fibers will not melt, but will decompose at about 500 °C (930 °F).

TABLE 2.2-6.   Kevlar Aramid Fiber Properties   (Reference 2-16)

	Kevlar	Kevlar 29	Kevlar 49
Density, lb./in.^{3}	0.52	0.52	0.52
Tensile strength, 10^3 psi	400	400	525
Tensile modulus, 10^6 psi	9	9	18
Tensile elongation, %	3 to 4	3 to 4	2 to 8
Coefficient of thermal expansion, in./in./°F			
Longitudinal Direction			
32 to 212°F	-1.1×10^{-6}		
212 to 392°F	-2.2×10^{-6}		
392 to 500°F	-2.8×10^{-6}		
Radial Direction			
32 to 212°F	33×10^{-6}		

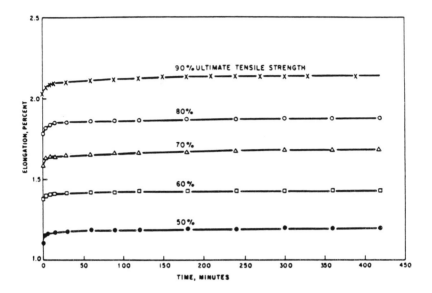

Figure 2.2-6. <u>The Effect of Stress on the Creep of Kevlar 49</u>
              (Reference 2-17)

TABLE 2.2-7.   The Effect of Chemicals on the Tensile Properties of Kevlar 49
Aramid   24-h exposure[a] (Reference 2-18)

Chemical	Tensile Strength, psi	Tensile Modulus, psi $\times 10^{-6}$
None (control)	411 000	18.33
Acetic acid (99.7% CH_3COOH)	431 600	18.16
Formic acid (HCOOH)	361 900	17.99
Hydrochloric acid (37% HCl)	419 200	17.80
Nitric acid (70% HNO_3)	165 200	17.40
Sulfuric acid	too weak to test	
Ammonium hydroxide (28.5% NH_3)	423 800	17.91
Potassium hydroxide (50% Solution)	305 900	17.69
Sodium hydroxide (50% Solution)	369 500	17.45
Acetone	423 100	18.22
Benzene (C_6H_6)	420 900	17.91
Carbon tetrachloride (CCl_4)	422 000	18.46
Dimethylformamide (DMF)	418 600	17.97
Methylene chloride (CH_2Cl_2)	425 900	18.30
Methyl ethyl ketone (MEK)	424 600	17.98
Trichloroethylene ("Triclene")	404 700	18.17
Chlorothene (1,1,1-trichloroethane)	418 600	18.32
Toluene ($C_6H_5CH_3$)	413 600	18.27
Benzyl alcohol ($C_6H_5CH_2OH$)	412 300	18.08
Ethyl alcohol (CH_5OH)	417 000	18.02
Methyl alcohol (methanol)	407 500	17.90
Formalin (HCHO)	405 500	17.87
Gasoline (regular)	419 900	18.37
Jet Fuel (Texaco "Abjet" K-40)	393 400	18.09
Lubricating oil ("Skydrol")	422 700	18.08
Salt water (5% solution)	410 100	16.92
Tap water	417 200	18.27

[a] Yarns were tested using air-actuated 4-C cord and yarn clamps on an Instron test machine, at 10 in. gage length with 3 turns per inch twist added, 10 percent per minute elongation, and at 55 percent relative humidity and 72°F.
Conversion factor—MPa = lb/in.$^2 \times 6.895 \times 10^{-3}$.

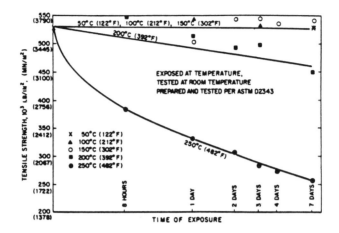

Figure 2.2-7.   The Effect of Elevated Temperature Exposure on the
Tensile Strength of Kevlar 49/Epoxy Roving Strands
(Reference 2-19)

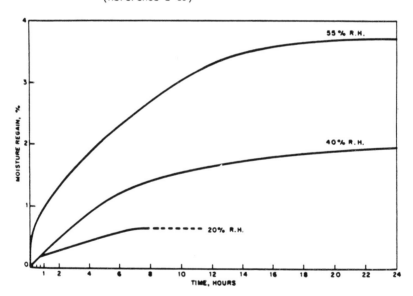

Figure 2.2-8.   The Effect of Relative Humidity on the Equilibrium Moisture
Regain and Regain Rate of 380 Denier Yarn of Kevlar
49 Aramid (Reference 2-20)

2.2.4 Silicon Carbide Reinforcements. Because of their compatibility with molten aluminum, their high temperature stability, and their low cost, silicon carbide fibers have become desirable components of advanced metal matrix composites. Silicon carbide fibers are produced by vapor deposition on either a tungsten or carbon filament substrate. This is done by heating the substrate and running it through a hydrogen/silane environment. Table 2.2-8 describes various properties of silicon carbide fibers, which are comparable to that of boron fibers. Silicon carbide fibers retain usable strengths at up to 1000 °C (1900 °F), and they have the unique characteristic of retaining basic fiber strength in molten aluminum for up to 30 minutes, as described in Figure 2.2-9. This characteristic greatly reduces the cost and difficulty of producing silicon carbide-aluminum composites. Silicon carbide (SiC) filament is a potentially low cost/high performance filament suitable for many advanced metal matrix applications.

TABLE 2.2-8. Silicon Carbide Filament Room Temperature Properties
(Reference 2-21)

Diameter, in.	0.0056
Ultimate tensile strength, 10^3 psi	485+
Tensile modulus, 10^6 psi	58
Shear modulus, 10^6 psi	24
Coefficient of thermal expansion, in./in./°F ($\times 10^6$)	2.7
Density, lb/in.^{3}	0.111
Length/lb, ft	30 500

Conversion factors—
10^3 psi = 6.9 MPa,
10^6 psi = 6.9 GPa,
in./in./°F = 0.556 cm/cm/°C,
lb/in.^{3} = 0.036 $\times$ g/cm^3,
1 ft = 0.305 m, and
1 lb = 0.45 kg.

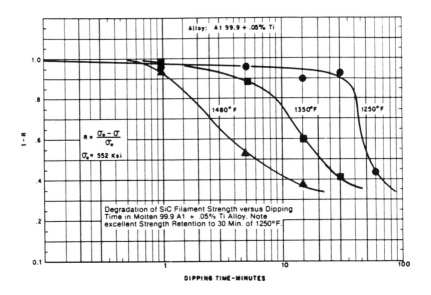

Figure 2.2-9. SiC Strength Versus Temperature and Time of Exposure to Molten Aluminum (Reference 2-22)

2.2.5. Fiberglass Reinforcements. The most common reinforcing fiber for polymer matrix composite materials is fiberglass. Table 2.2-9 compares the properties of two common glasses and quartz to that of other fibers. The advantages of fiberglass include low cost, high tensile and impact strengths, and high chemical resistance. The disadvantages include low modulus, self-abrasiveness, low fatigue resistance, and poor adhesion to matrix resins. Quartz fibers are produced from natural quartz crystals and are used when dielectric properties and high temperature resistance are important.

Molten glass is produced from melting a dry mixture of sand, limestone, boric acid in a high refractory furnace. This molten glass is either poured directly into the fiber drawing furnace or made into marbles and remelted later. Continuous fibers are produced by mechanically attenuating droplets of molten glass. Discontinuous fibers are made by blowing jets of air across an orifice at the bottom of the molten glass, forming filaments 20 to 38cm long. These are collected and gathered into a strand. Other unique properties can be found in Table 2.2-10 and Figure 2.2-10.

TABLE 2.2-9. Comparative Properties of Glass Type Fibers (Reference 2-23)

Type of Fiber	E-Glass	S-Glass	Quartz	Boron	Graphite Type "A"	Kevlar 49
Fiber density						
lb/in.^{3}	0.092	0.090	0.079	0.079	0.063	0.052
g/cm^3	2.54	2.49	2.2	2.68	1.74	1.44
Tensile strength						
10^3 psi	500	665	450	500	250 to 300	525
MPa	3448	4585	3100	3448	1724 to 2069	3620
Modulus of elasticity						
10^6 psi	10.5	12.4	10	60	24/32	19
GPa	72.4	85.5	68.9	413.7	172.4 to 220.6	131.0
Range of diameters						
mils	0.1 to 0.8	.3 to .7	6.4	4 to 8	0.35 to 0.5	0.47
mm	0.003 to 0.020	0.008 to 0.013	0.010	0.10 to 0.20	0.009 to 0.013	0.0119
Coefficient of thermal expansion						
in./in./°F × 10^{-6}	2.8	1.6	1.6	2.8	−0.6	−1.1
cm/cm/°C × 10^{-6}	5.0	2.9	2.9	5.0		−2

TABLE 2.2-10. Fiberglass Filament Designations (Reference 2-24)

Filament Designation	Filament Diameter	
	in.	μm
B	1.5×10^{-4}	3.8
C	1.8×10	4.5
D	2.1×10	5
DE	2.5×10	6
E	2.9×10	7
G	3.6×10	9
H	4.2×10	10
K	5.1×10	13

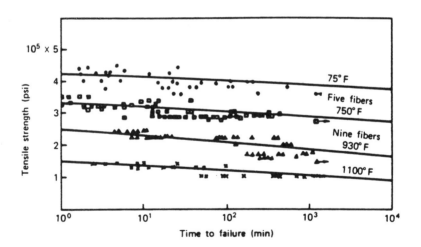

Figure 2.2-10. Static Fatigue Test Results for E-Glass Fibers (Reference 2-25)

2.3 Manufacturing Processes for Composite Structures.

2.3.1 Fabrication Methods for Metal Matrix Composites. Metal matrix composites are produced in two basic steps. The first step is the creation of the metal matrix composite from its composites. The form is usually not in its final state after this step, and a second step is required to finish by machining, rolling, forming, or metallurgical bonding.

A major consideration of the metal matrix formation is the reaction of the reinforcing fibers to the molten metal. This may limit the use of many fibers, but a protective coating may be put on the fibers to facilitate their use.

One fabrication procedure involving larger diameter fibers, such as boron and silicon carbide fibers, creates a monolayer tape by hot pressing a layer of fibers between two foils. The metal flows around the fibers, and diffusion bonding occurs. The same method is employed to make laminates with fibers oriented in a specific direction.

Composite materials may also be produced by infiltrating molten metal into fabric of preform. The preform fibers are usually held together by ceramic or organic materials which burn away during infiltration. This process may be done under pressure, in a vacuum, or both. This is the most common method for producing graphite/aluminum and graphite/magnesium composites. The procedure may sometimes involve the intermediate step of producing a bundle of infiltrated fibers called a "wire." The wire is then formed into final shape using hot press techniques similar to that of the larger fibers. Direct casting is used when an "air stable" coating is used on the fibers.

Whisker and particulate reinforcements are usually mixed with powdered metal to form composites. A liquid metal method is used, but details are proprietary. The actual mixing takes place either in a ball mill or with the aid of a liquid, which is removed later. Mixtures are hot pressed into billets and then formed using rolling, extrusion, spinning, forging, or creep-forming. Machining may also be used, but may present difficulties because of the hardness of reinforcements.

2.3.2. Fabrication Methods for Polymer Matrix Composites. Polymer matrix composites can be produced by many different methods including wet layup, filament winding, compression molding, autoclave molding, injection molding, and pultrusion.

In the wet layup method, precut materials are fitted into a female mold. Resin is then applied, worked in, and allowed to cure at room temperature. Some materials such as Kevlar are not available for this fabrication method.

The filament winding technique involves winding fibers on a slowly turning mandrel which is the shape of the inside of the desired form. The winding process may be hand fed, but is usually electronically controlled. Winding patterns may vary from nearly parallel to nearly perpendicular to the axis of rotation. The entire assembly is cured and the mandrel is removed. Machining may be necessary to achieve the desired finished dimensions. The advantages of this method include the ability to place large amounts of composite precisely with little manpower, automated processing, repeatable processing, and low scrap rate. The disadvantages include low longitudinal strength and stiffness, and the expense of producing and removing mandrels, especially for complex parts.

In the compression molding technique, the fiber and resin are placed within a metal cavity, pressurized, and heated until cured. The material will then have a rigid shape in the form of the cavity. Either continuous or chopped fibers may be used in this method, or a combination of the two may be used. If continuous fibers are used, they must be preformed to closely match the cavity shape.

Autoclave molding is used for molding thermoset preimpregnated materials. This method provides low cost fabrication of complex parts. Starting materials include prepreg tape, woven cloth, or mat. The process itself involves laying up the prepreg laminates on a mold, covering them with flexible sheets, and vacuum sealing. The assembly is then cured at a specified heat and pressure in a vacuum. The process is inexpensive and is most commonly used for limited production run parts.

Injection mold processing consists of heating the materials in a barrel and transferring the flowable material into a cooler metal cavity with a plunger. The materials are rapidly cooled and ejected in the form of the final part. Two thirds of the plungers used are reciprocating screw type, the other third are ram plungers. The advantages of this method include high production rates, low labor cost, compatibility to small parts, and precise dimensions with little finishing. The major disadvantage of this method is that continuous reinforcements cannot be used.

Pultrusion continuously produces constant cross-section continuous parts. A continuous yarn of carbon, glass, or Kevlar is resin coated and pulled through a heated forming die. The resin sets during passage through the die. This process is generally limited to thermoset resins such as epoxies and polyesters. The advantages of this method include almost no scrap loss, low labor costs, and high strength and stiffness. The disadvantages include the restriction to constant cross-section parts and difficulties in using mat fabric and industrial fibers.

3. Metal Matrix Composites

3.1 <u>Introduction</u>. In metal matrix composites, various mechanical properties are superior to their unreinforced monolithic metal counterparts. Specifically, this is true in the areas of high temperature stability, specific strength, and modulus. As a result, metal matrix composites offer promise for many different aerospace applications. In Table 3.1-1 some of the advantages of metal matrix composites over resin matrix composites in different applications are shown.

Because of low temperature and low cost fabrication methods, resin matrix composite development has remained ahead of the higher temperature, higher cost fabrication methods of metal matrix composites. However, there are three important general classes of metal matrices that are reinforced by the high-performance fiber materials. These are aluminum, magnesium, and titanium. Table 3.1-2 provides some properties of these metal matrices for advanced metal composites.

Aluminum and magnesium matrices have advantages over polymer-matrix composites because of their higher melting temperatures and lower densities. As a result of oxidation and corrosion considerations, aluminum is a better metal matrix than magnesium even though it is more dense. Because of these considerations, the discussion will focus primarily on aluminum matrix composites.

Aluminum alloy metal matrix composites that contain boron or Borsic SiC coated fibers have been the most extensively studied and developed metal-matrix composite system. As a result of carbon fibers being less expensive, carbon fiber-reinforcement is of continuing interest. Other composites such as magnesium that can be based on these fibers are also being investigated. Silicon carbide and aluminum oxide are other examples of reinforcements for magnesium composite systems.

Because of its material properties such as higher temperature melting point, there has been substantial interest in titanium-matrix composites.

66

TABLE 3.1-1.   Metal-Matrix Material Characteristics (Reference 3-1)

Key characteristics of metal matrix composite (compared to plastic matrix composite)	Potential advantages in space applications
Greater matrix strength and modulus possible (a) tensile (b) shear (c) compressive (d) bearing	Compared to uniaxial plastics: More efficient in plate buckling More efficient for transverse and off-axis loads More efficient in combined loads of tension, compression, or shear Less tendency for matrix crazing
Harder surface	Greater erosion resistance (potential nozzle or leading-edge applications)
Higher-temperature matrix capabilities	Higher temperature applications for uniaxial material, particularly above 600 F, where plastics are limited
Greater thermal conductivity	Avoids local hot spots by spreading out heat, operates at lower temperature; better for nozzles, flame impingement surfaces, and leading-edge applications
Higher electrical conductivity	Better lightning protection (aircraft, boosters) Bleeds off static electricity, avoiding explosion of fuels (aircraft, boosters) Reduces electromagnetic interference problems
Matrix is inorganic	Better behavior for spacecraft applications where flammability (in high O_2 atmospheres), lower temperatures (-300 F), and radiation may result in problems with plastics
Available in sheet form	More conventional methods of manufacturing possible
Post-buckling shear strength	More efficient shear-skin panels can be designed (since buckling can be tolerated)

TABLE 3.1-2.   Properties of Aluminum, Magnesium, and Titanium Matrix Metals (Reference 3-2)

Base Metal	Alloy	Density, g/cm^3	Modulus, psi	Ultimate Tensile Strength, psi	Thermal Expansion Coefficient $\times 10^{6}$, 70 to 200°F
Aluminum	Alloy 1100	2.71	10 000 000	12 000 (annealed)	13.2
		2.71	10 000 000	13 000 (annealed)	13.1
	Alloy 2024	2.77	10 600 000	27 000 (annealed)	12.9
				65,000 (T-3)	
	Alloy 5052	2.69	10 200 000	28 000 (annealed)	13.2
				38 000 (H-34)	
	Alloy 5056	2.63	10 300 000	42 000 (annealed)	13.4
				60 000 (H38)	
	Alloy 6061	2.71	10 000 000	18 000 (annealed)	13.0
				45 000 (T-6)	
	Alloy 7075	2.80	10 400 000	33 000 (annealed)	13.1
				83 000 (T-6)	
	Alloy 201	2.80	...	65 000 (solar treated and aged)	10.7
Magnesium	AM100A	1.74	6 300 000	40 000 (T-6)	11.9
	AZ63A	1.80	6 500 000	27 000	14.5
				40 000 (T-6)	
	EZ33A	1.82	6 500 000	34 000 (T-6)	14.5
				20 000 (T-5)	
	HK31A	1.83	6 500 000	29 000 (annealed)	14.5
		1.79		34 000	14.5
Titanium	commercial Ti	4.5	16 200 000	80 000	...
	Ti-6Al-4V	...	...	170 000 (aged)	...
	Ti-4Al-4Mn	...	...	185 000 (aged)	...
	Ti-7Mn	...	...	150 000	...
		...	...	...	...

Even though titanium has a greater density than magnesium and aluminum, it is still a high temperature, low density material with good mechanical properties. In general, titanium matrix is very reactive. It does not form protective oxide barriers on its surface. Therefore, in most cases the melt infiltration fabrication method proves to be unusable, and possible combinations of materials are restricted by fiber matrix interactions. Silicon carbide, silicon carbide-coated boron, and boron have been the reinforcements of the greatest importance.

High operating temperatures for metal matrix composite materials reactively alter the strength of the composite, and a great amount of work is being done to lessen or overcome these adverse effects.

3.2 <u>Aluminum Metal Matrix Composites.</u>

3.2.1 Background. The aluminum series 1000, 2000, 5000, 6000, and 7000 are generally the types of aluminum matrix materials employed in advanced composites. To date, the aluminum alloys 2024, 2219, 6061, 7090, and 7091 are most often used with different temperatures. Table 3.2-1 shows each series with its principal alloying elements. Displayed in Table 3.2-2 are the physical and mechanical properties of commonly used aluminum alloys. Each alloying element imparts distinct properties to the aluminum. The most widely and best developed aluminum matrix composite systems are those made using reinforcement with graphite, silicon carbide, and boron fibers.

Because of their low graphite cost, matrix attainability, better per-formance as well as decreased manufacturing and fabrication expenses as compared with other composite systems, graphite/aluminum composites have higher long term possibilities. The DoD has widely funded programs for development of graphite/aluminum composites since the late 1960's.

The composite system that has been suggested more than any other in structural applications is the boron and borsic/aluminum composite system. This fiber-reinforced, high modulus, metal matrix composite system is the most developed and has the most extensive use. Also under development, silicon carbide/aluminum composites have great possibilities for commercial applications dealing with cost and physical properties.

TABLE 3.2-1.   Aluminum Matrix Series and Their Principal Alloying Element
               (Reference 3-3)

Alloy Series	Alloying Element
1xxx	None
2xxx	Copper
3xxx	Manganese
4xxx	Silicon
5xxx	Magnesium
6xxx	Magnesium Silicide
7xxx	Zinc

TABLE 3.2-2.   Mechanical and Physical Properties of Aluminum Matrix
               Alloys (Reference 3-4)

Property	Alloys			
	1100	2024	6061	7075
Solidus ($^\circ$C)	643	554	582	477
Liquidus ($^\circ$C)	657	649	652	635
Density (gm/cm^3)	2.71	2.75	2.70	2.80
Elastic Modulus (msi)	10.0	10.2	10.0	10.3
Specific Modulus[a] (msi)	3.69	3.71	3.70	3.68
Yield Strength at 0 (ksi)	5.0[b]	14.0	8.0	15.0
Yield Strength at T6 (ksi)	5.0[b]	46.0	40.0	73.0

[a]Specific modulus is E/ρ where is the elastic modulus and ρ is the density.

[b]1100 Al is not heat treatable and the yield strength increases with work hardening to the maximum value.

3.2.2 Properties of Aluminum Matrix Composites.

3.2.2.1 Graphite-Reinforced Aluminum. The consolidation process greatly affects the longitudinal properties in graphite/aluminum compos-ites. The chemical interplay between the matrix and fiber during consoli-dation possibly causes large drops in the strength of the consolidation plates.   This was discovered after consolidation by an increase in the aluminum carbide (Al_4C_3) compound at the interface.

Production for rayon based carbon fiber is ending even though it has the most congruent actions in dispersion and composite properties in the aluminum matrix. Basically, it is being discontinued because of its high cost.   Therefore, it cannot be supported as a viable means for the reinforcement of aluminum.

When exposed to molten aluminum alloys, PAN based carbon fibers are reactive and somewhat unpredictable. For example, in T-300 fiber reduction and low precursor wire, strengths result from exposure to molten aluminum alloys when HM-300 produces high strength precursor plates and wires. This could possibly be caused from the different exterior features of each fiber.

The desirable transverse strength for a typical metal matrix composite is 137.9 MPa. Using additional titanium and boron in the aluminum alloys increases the longitudinal graphite/aluminum properties, but does nothing for the crosswise properties. Graphite/aluminum composites have low cross-wise properties and low crosswire strengths.   The rayon based graphite/alloy 201 composite has a crosswise strength of 34.5 MPa, which is much less than 137.9 MPa. Graphite/aluminum composites have considerably low transverse strengths.

Corrosion resistance has been studied in graphite/aluminum composites. Galvanic corrosion is present in graphite/aluminum composites because of the galvanic couples of graphite and aluminum that exist, and also because couples are present in the aluminum matrix and strengthening foils. At low stress levels, fracture tracks and galvanic action can cause early failure. The unique physical and mechanical properties of graphite-reinforced aluminum can be found from Figures 3.2-1 to 3.2-7 and from Tables 3.2-3 to 3.2-5.

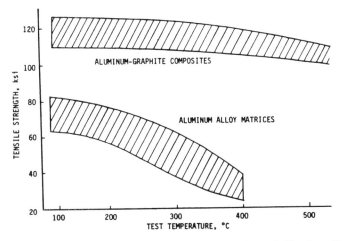

Figure 3.2-1. Typical Elevated Temperature Properties of Aluminum/Graphite Composites (Reference 3-5)

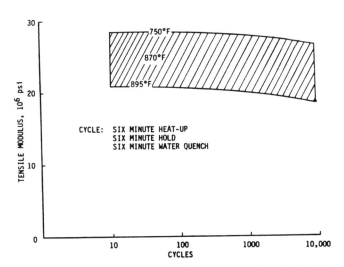

Figure 3.2-2. Typical Retained Room Temperature Tensile Modulus of 32 (v/o) T50 Graphite/Aluminum Composite Wire after Thermal Cycling (Reference 3-6)

TABLE 3.2-3. Tensile Properties of Various Aluminum Alloy-- Thornel-75 (Graphite) Composites (Reference 3-7)

Matrix composition	Specimen condition	Volume percent fiber	Strength					Average modulus	
			Average		Number of samples	Low value (psi)	High value (psi)	(GN/m²)	(psi)
			MN/m²	psi					
Commercially pure aluminum	As-infiltrated	32	68	99,000	8	65,000	116,000	178	25.7
	Pressed	35	65	95,000	7	85,000	104,000	147	21.3
Aluminum-7 w/o zinc	As-infiltrated	32	71	103,000	7	59,000	132,000	166	24.1
	Pressed	38	87	126,000	10	102,000	155,000	190	27.5
Aluminum-7 w/o magnesium	As-infiltrated	31	68	98,000	4	87,000	124,000	195	28.1
Aluminum-13 w/o silicon	As-infiltrated	22	55	80,000	7	73,000	88,000	165	23.8

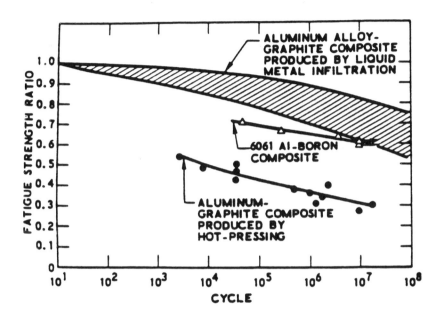

Figure 3.2-3.   Comparison of Fatigue Behavior between Aluminum-Graphite Composites and Other Metal Matrix Composites (Reference 3-8)

TABLE 3.2-4.   Corrosion Behavior of Aluminum--Thornel-50 (Graphite) Composite for 1000 Hr (Reference 3-9)

Environment	356 aluminum		356 aluminum–25 v/o Thornel-50	
	$(23°C)$	$(50°C)$	$(23°C)$	$(50°C)$
Distilled water	Nil	Nil	1.2	1.2
3.5 % NaCl solution	1.1	4.9	4.7	9.8

[a]Mils per year (1 mil = 25 μm).

TABLE 3.2-5.   Summary of Transverse Tensile Strengths of Various Aluminum--Graphite Composite Systems (Reference 3-10)

Composite		Average		High (psi)	Low (psi)	Number of tests
Fiber	Matrix	(MN/m^2)	(psi)			
Thornel-50	Al–12Si	26	3777	6500	433	9
Courtaulds	220 Al	42	6117	8690	3760	20
Courtaulds	356 Al	70	10,008	14,600	5500	26
Courtaulds HM	Al–10Mg	29.5	4280	4500	3600	5
Whittaker–Morgan	356 Al	50	7300	11,300	4100	5
Whittaker–Morgan	7075 Al	21	3040	5100	400	5

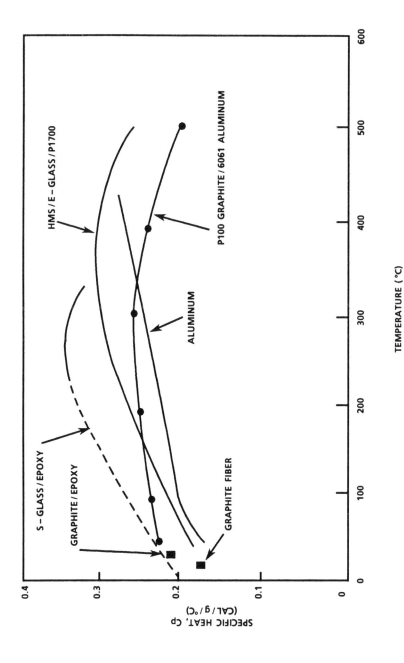

Figure 3.2-4.   Specific Heat of Graphite/Aluminum Composite (Reference 3-11)

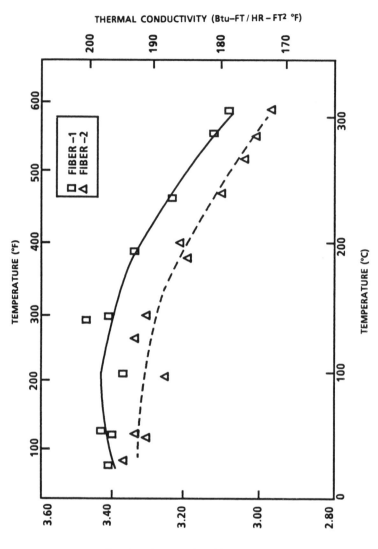

Figure 3.2-5.   Thermal Conductivity P100 Graphite/6061 Aluminum Composite
(Reference 3-12)

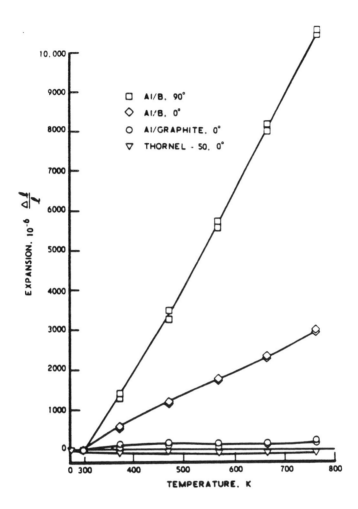

Figure 3.2-6. Thermal Expansion versus Temperature for Graphite/Aluminum
(Reference 3-13)

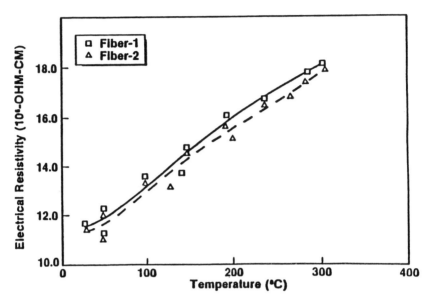

Figure 3.2-7. Electrical Resistivity of Graphite P100/6061 Aluminum Composite
(Reference 3-14)

3.2.2.2 Boron-Reinforced Aluminum. Borsic/aluminum and boron/alumi-
num composites can display tensile strengths greater than 1379 MPa. This
is due to high strength boron fibers when the borsic/aluminum composite
nears the rule of mixture levels. The size of the filament's diameter
affects the properties of the composite. It is often found that the
smaller the filament diameter, the weaker the composite properties (Table
3.2-6).

Graphite/aluminum composite transverse properties are inferior compared with boron/aluminum and borsic/aluminum composite properties. In
Table 3.2-7 common values of transverse properties for some boron/aluminum
and borsic/aluminum systems are shown. A composite with 45 percent boron
in 2024-T6 alloys displays transverse strengths of 331 MPa while a
composite with 50 percent boron in 6061F alloys normally has transverse
tensile strengths of 124.1 MPa. The average transverse strength is
dependent upon the quality and the matrix of the alloy.

When compared, the longitudinal strengths of these composites are on the order of 10 to 20 percent greater than the crosswise strengths. As a result of this large difference, boron/aluminum and borsic/aluminum composites are often produced with the filaments arranged in random directions.

Fatigue loading in boron/aluminum composites is superior, and is valuable in system for structural applications. Fatigue strength is dependent upon many things, including work hardening, tensile strength, modulus, notch sensitivity as well as chemical reaction properties and products. At 60 percent volume, boron/aluminum composites at $x10^7$ cycles have obtained fatigue strengths of 1379 MPa. The reaction of composite materials can also be affected by the modulus and strength of the filament matrix bond as well as previous or consolidation-generated defects. Other unique physical and mechanical properties of boron-reinforced aluminum can be found from Figures 3.2-8 to 3.2-15 and from Tables 3.2-7 to 3.2-9.

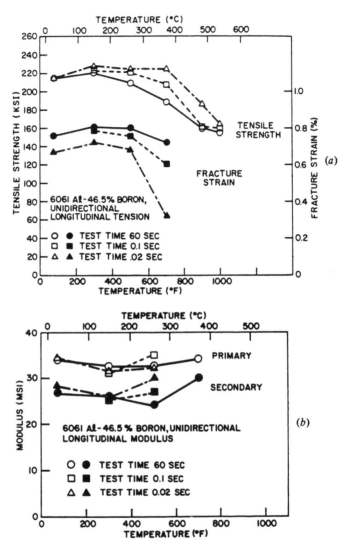

Figure 3.2-8.   Tensile Properties of 6061 Aluminum-46.5 Percent Boron
Unidirectional Composite at Several Temperatures and
Strain Rates. (a) Longitudinal Ultimate Strength and
Fracture Strain; (b) Longitudinal Young's Moduli
(Reference 3-17)

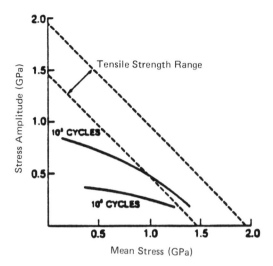

Figure 3.2-9. <u>Fatigue Failure Envelopes for Boron-Aluminum</u>
(Reference 3-18)

TABLE 3.2-8. Maximum Temperatures for Metals Reinforced with
Boron and Carbon Fibers (Reference 3-19)

System	Temperature (C)	Remarks
C-Al	500	Al contains 12% Si
B-Al	540	B coated with SiC
B-Ti	650	B coated with SiC
B-Ti	540	Oxygen present; B coated

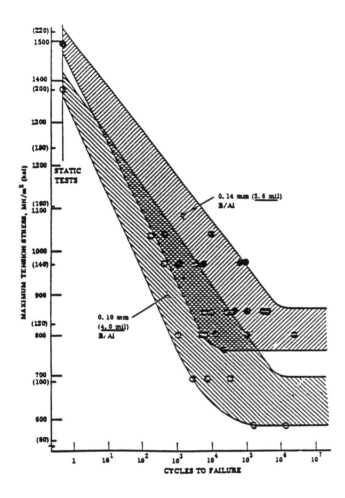

Figure 3.2-10.　Fatigue Properties of B/Al Composite 50 v/o
Unidirectionally Reinforced, R = -0.1
(Reference 3-20)

TABLE 3.2-6. Longitudinal Tensile Properties of B/Al Composites
(Reference 3-15)

Matrix	Fiber, volume %	Fiber Type and Diameter	Ultimate Tensile Strength, psi	Modulus, psi
2024	47	boron, 5.6 mil	212 000	32 000 000
6061	50	boron, 5.6 mil	208 000	32 000 000
6061	54	boron, 5.7 mil dia	203 000	36 600 000
6061	30	boron, 5.7 mil dia	115 000	17 600 000
6061	50	boron, 4.0 mil	180 000	...
6061	50	boron, 8.0 mil	210 000	...

TABLE 3.2-7. Typical Transverse Properties of B/Al and Borsic/Al Composites
(Reference 3-16)

Matrix	Boron, volume %	Borsic, volume %	Transverse Strength, 10^3 psi	Modulus, 10^6 psi
2024 F	45 (5.6 mil)		27	...
2024-T6	45 (5.6 mil)		48	...
6061 F	50 (5.6 mil)		18.9	...
6061-T6	50 (5.6 mil)		41.7	...
6061 F		52 (5.7 mil)	20	...
6061-T6		52 (5.7 mil)	36	19
2024 F		56 (5.7 mil)	22	23
2024-T6		56 (5.7 mil)	46	25

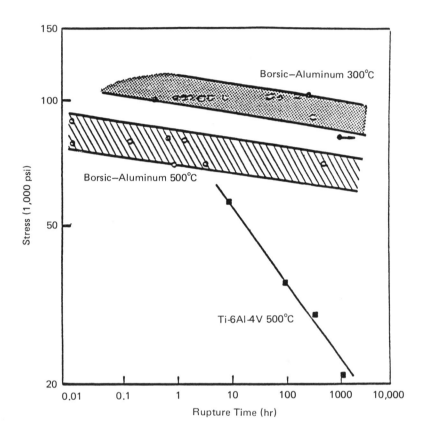

Figure 3.2-11. Stress Rupture Properties of Borsic-Aluminum Composites
50 Percent Fiber by Volume (Reference 3-21)

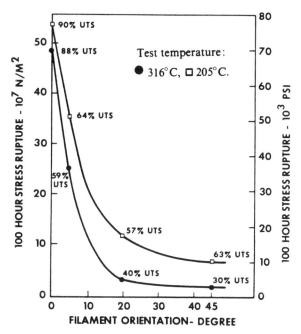

Figure 3.2-12.   Applied Tensile Stress to Cause Rupture at 100 Hr as a
                 Function of Angle to Principal Fiber Axis for Aluminum
                 Reinforced with 100μm Boron (Reference 3-22)

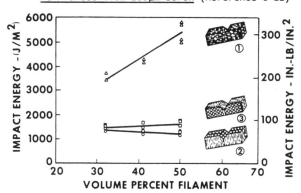

Figure 3.2-13.   Notched Charpy Impact Energy as a Function of Specimen
                 Orientation and Volume Fraction of Fiber. Material is
                 6061 Aluminum Reinforced with 100-μm Borsic
                 (Reference 3-23)

TABLE 3.2-9.   Notched Charpy Impact Energy at 23 °C for Boron/Aluminum Composites Containing 50–60 Percent Fiber by Volume (Reference 3-24)

Spec #	Matrix*	Fiber	Fabrication Conditions			Orientation	Impact Energy	
			Temp °C	Pressure 10^6 N/m²	Time min		J	Ft-lbs
1	6061-F	BORSIC	565	145.0	60	LT	6.7	5.0
2	6061-F	BORSIC	565	145.0	60	TT	2.0	1.5
3	6061-F	BORSIC	565	145.0	60	TL	1.3	1.0
4	6061-F	BORSIC	490	72.5	30	LT	7.8	5.8
5	6061-F	BORSIC	450	72.5	30	LT	9.4	7.0
6	6061-F	Boron	450	145.0	30	LT	17.7	13.1
7	6061-F	Boron	480	145.0	30	LT	13.2	9.7
8	6061-F	Boron	450	72.5	30	LT	18.5	13.7
9	1100	BORSIC	450	72.5	30	LT	18.4	13.6
10	1100	Boron	450	72.5	30	LT	26.0	19.2
11	1100	Boron	450	72.5	30	LT	22.8	16.9
12	1100	Boron	450	72.5	30	LT	28.4	21.0
13	1100	Boron	450	72.5	30	LT	30.0	22.2
14	1100	Boron	450	72.5	30	LT	>30.0	>22.3
15	1100	Boron	450	145.0	30	LT	26.1	19.3
16	1100	Boron	450	145.0	30	LT	21.5	15.9
17	1100	Boron	450	145.0	30	LT	28.2	20.9
18	2024-F	Boron	450	72.5	30	LT	8.1	6.0
19	2024-F	Boron	450	72.5	30	LT	15.4	11.4
20	5052/56	Boron	450	145.0	30	LT	8.0	5.9
21	6061/1100-F	Boron	450	145.0	30	LT	26.6	19.7
22	6061/1100-F	Boron	450	145.0	30	LT	25.6	18.9
23	6061/1100-T6	Boron	450	145.0	30	LT	22.3	16.5

*F and T-6 indicate the as-fabricated and heat-treated conditions respectively

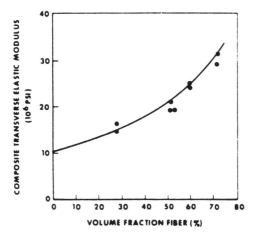

Figure 3.2-14. Transverse Tensile Elastic Modulus for 5.7 mil Borsic-6061 Composites (Reference 3-25)

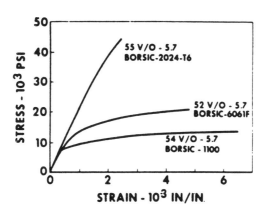

Figure 3.2-15. Transverse Tensile Stress-Strain Curves for Borsic-2024 Composites (Reference 3-26)

3.2.2.3 Silicon Carbide-Reinforced Aluminum. Silicon carbide aluminum composite stands to be a competitor of boron-aluminum systems because
of high environmental resistance and mechanical properties such as its
ability to mix well with molten aluminum, high temperature reliability, and
low raw material cost.   As a result silicon carbide (SiC) filament is
possibly a high performance/low cost metal matrix filament that can be used
in composite structural applications.   In an attempt to keep cost down, die
casting and extrusion fabrication methods have been employed in the
production of the aluminum/SiC whisker composites.

SiC whiskers have a high tendency to align when directed forces like
forging, rolling and extrusion are applied to the composite. They align in
the direction of the applied load. Because of their small size, desirable
orientation is gained.   This amplifies the aluminum alloy's longitudinal
properties.   The whiskers can also be wetted and bonded easily to the
aluminum alloy.   The composites also have very appealing crosswise
properties.   For example, 344.7 MPa is the lowest crosswise strength.
Crosswise modulus was measured with values of 103.4 GPa.   The parts that
were measured had undergone high extrusion.   The ratio of extrusion was 30
to 40.   A 2 in. (50 mm) diameter billet was extruded to a .2 in. (.5 mm)
sheet. Tubes can also be extruded.

Al/SiC whiskers components impact and corrosion characteristics have
been evaluated in introductory studies.   The results obtained have not
uncovered any problems. Other unique physical and mechanical properties of
silicon carbide reinforced aluminum can be found in Figures 3.2-16 through
3.2-27 and in Table 3.2-10.

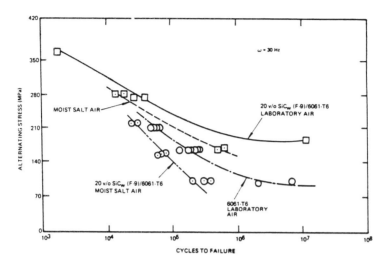

Figure 3.2-16.   Alternating Stress versus Number of Cycles for
20 v/o SiC$_W$ (F-9)/6061-T6 Al Specimens Compared to
6061-T6 Subjected to Corrosive Environments
(Reference 3-27)

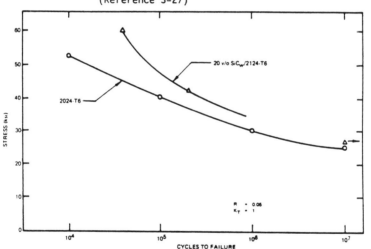

Figure 3.2-17.   Room Temperature Fatigue Data for Unreinforced and Reinforced
2124-T6 Al Composite (Reference 3-28)

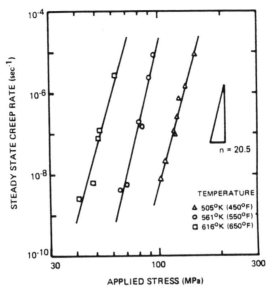

Figure 3.2-18. Steady State Creep Rate as a Function of Applied Stress for 20 w/o SiC$_w$/6061 Al at Temperatures between 505° and 616°K (Reference 3-29)

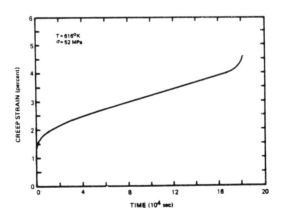

Figure 3.2-19. Typical Creep Curve for 20 w/o SiC$_w$/6061 Al Composite (Reference 3-30)

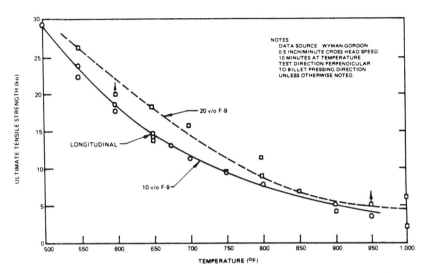

Figure 3.2-20. Ultimate Tensile Strength versus Temperature for SiC_w (F-9)/2124 Al (SXA™ 24) Billet (Reference 3-31)

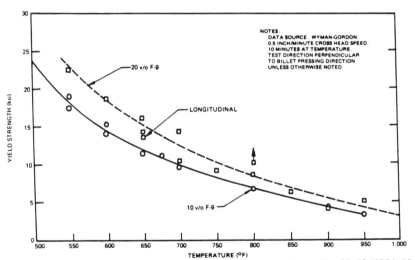

Figure 3.2-21. Yield Strength versus Temperature for SiC_w (F-9)/2124 Al (SXA™ 24) Billet (Reference 3-32)

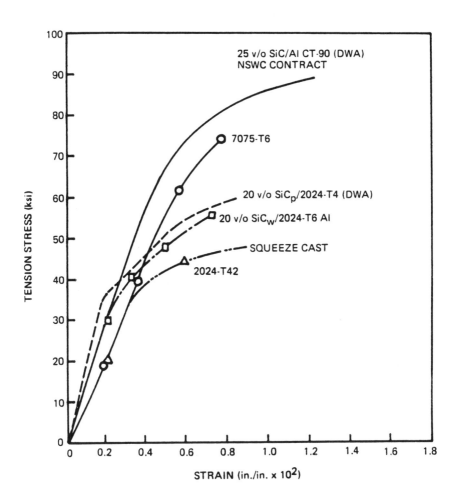

Figure 3.2-22.   <u>Stress-Strain Curves for a Number of SiC/Al Composites</u>
                 (Reference 3-33)

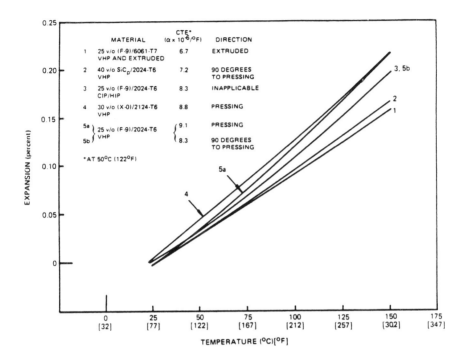

Figure 3.2-23.   **Linear Coefficient of Thermal Expansion for Five SiC/ Aluminum Matrix Composites** (Reference 3-34)

TABLE 3.2-10.　Thermal Properties of SiC$_w$/2024 Al at Room Temperature (Reference 3-35)

Material	Conductivity, k (Btu/ft/sec °F)	Specific Heat, C$_p$ (Btu/lb$_m$ °F)	Density (lb$_m$/ft^3)
5083 Al	0.019	0.23	165.9
7075 Al	0.025	0.23	165.9
15.7 v/o SiC$_w$/2024 Al	0.0158	0.24	177.9
25 v/o SiC$_w$/2024 Al	0.0125	0.24	177.9

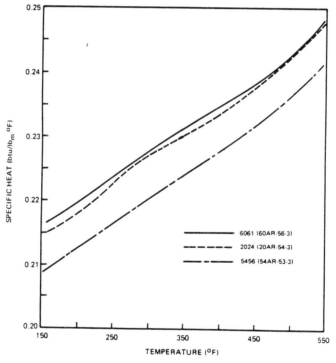

Figure 3.2-24.　Specific Heat Versus Temperature for 15 v/o SiC$_w$/Al (Reference 3-36)

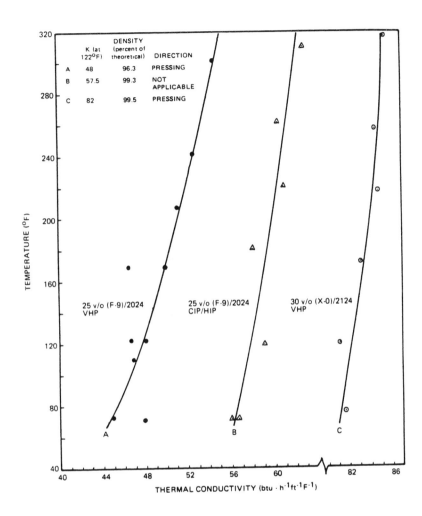

Figure 3.2-25.   Thermal Conductivity of Three SiC/Aluminum Matrix Composite
                 (Reference 3-37)

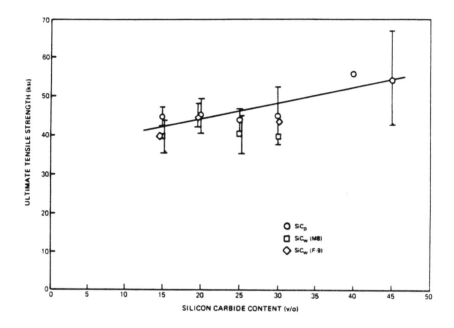

Figure 3.2-26.   Ultimate Strength as a Function of SiC Reinforcement Content
for as-Fabricated 2024 Matrix Composites (Reference 3-38)

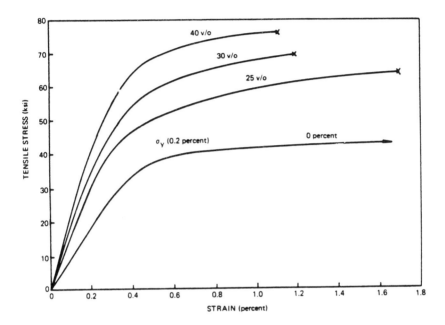

Figure 3.2-27.   Stress-Strain Curves for 6061-T6 Aluminum with Various
                 Levels of SiC_p Reinforcement (Reference 3-39)

3.3 Magnesium Metal Matrix Composites.

3.3.1. Background. Magnesium   is   an   excellent   candidate   for lightweight, high strength composite matrix material.   Advantages of magnesium matrices include a low density and low reaction rate between the matrix and fiber during fabrication.   Magnesium matrices have been used with brittle fibers such as boron, silicon carbide, and graphite filaments. Magnesium matrix composites typically have weights 30% lighter than aluminum alloys.   The specific properties of magnesium composites are excellent, and thermal expansion remains zero or near zero through a wide range of temperature.   This excellent thermal expansion is due to the combination of properties of the matrix and fiber, and like strength properties, may be tailored to meet specific application requirements.

Magnesium matrix composites are currently available in cast shapes including flat plates, T-sections, rods, hollow cylinders, and irregular specialty parts.   Production methods for new cast shapes must take into account the orientation and holding of the fibers during fabrication.   For complex systems, this may prove difficult and expensive.   Costs are expected to decrease as less laborious production techniques are developed. Initial production costs substantially more than subsequent castings.

3.3.2 Properties of Magnesium Matrix Composites. Magnesium is one of the few lightweight metals which does not offer problems in interaction with the advanced reinforcement fibers such as boron, silicon carbide, and graphite.   At the present time graphite, alumina, and boron fibers are being   strongly   considered   for   use   with   magnesium   matrix. Graphite/magnesium offers the unique advantages of near-zero thermal expansion over a wide range of temperatures.   Graphite reinforced magnesium with 40 percent by fiber volume possesses a typical tensile strength of 120,000 psi and an elastic modulus of about 25 million psi.   Unique physical and mechanical properties of advanced fibers reinforced magnesium matrix are summarized and presented from Figures 3.3-1 through 3.3-6 and from Tables 3.3-1 through 3.3-3.

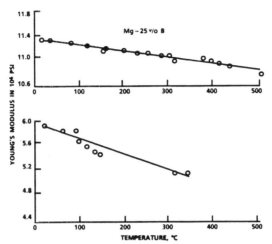

Figure 3.3-1.   Elastic Modulus of Magnesium Matrix With and Without
                Boron Filament as a Function of Temperature
                (Reference 3-40)

Table 3.3-1   Charpy Fracture Energy for Alumina/Magnesium
              (50 percent V_f) (Reference 3-41)

MATERIAL CONDITION	TEST TEMPERATURE	FIBER DIRECTION	TEST DIRECTION	CHARPY IMPACT ENERGY (FT – LBS)
AS RECEIVED	RT	0°	0°	0.43
		0°	90°	0.32
		±22¼°	0°	0.54
		±22¼°	90°	0.45
	300 F	0°	0°	0.49
		0°	90°	0.25
		±22¼°	0°	0.32
		±22¼°	90°	0.28
T5 (HEAT TREATED)	RT	0°	0°	0.44
		0°	90°	0.26
		±22¼°	0°	0.44
		±22¼°	90°	0.35
	300 F	0°	0°	0.38
		0°	90°	0.24
		±22¼°	0°	0.36
		±22¼°	90°	0.229

Table 3.3-2   Average Ultimate Tensile Strength and Elastic
              Modulus Properties of Boron-Magnesium Composites
              (Reference 3-42)

FILAMENT (PERCENT VOLUME)	U. T. S. (KSI)	ELASTIC MODULUS (MSI)
50	16.7	7.3
35	69.5	22.0
28	79.0	17.2
19	45.2	15.9
10	22.4	8.6
6	15.6	8.4

Table 3.3-3   Average Ultimate Tensile Strength and Elastic
              Modulus Properties of Tantatum-Magnesium
              Composites (Reference 3-43)

FILAMENT (PERCENT VOLUME)	U. T. S. (KSI)	ELASTIC MODULUS (MSI)
40	28.8	13.2
32	32.2	13.0
28	30.2	11.8
19	23.1	10.4
10	15.9	5.2
0	12.9	6.4

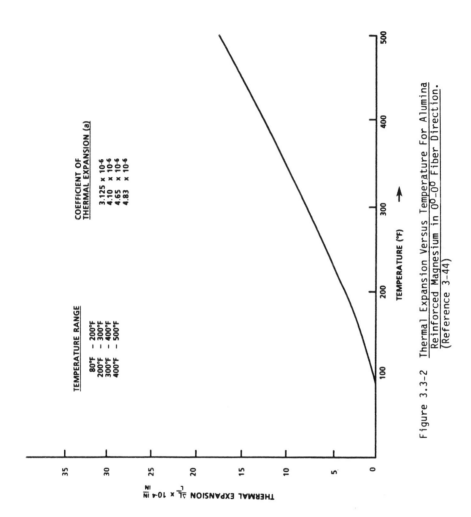

Figure 3.3-2   Thermal Expansion Versus Temperature For Alumina Reinforced Magnesium in 0°-0° Fiber Direction. (Reference 3-44)

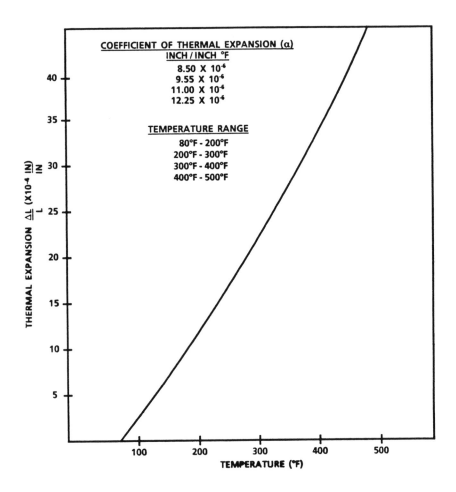

Figure 3.3-3   Thermal Expansion Versus Temperature For Alumina
Reinforced Magnesium in 0-90° Fiber Direction.
(Reference 3-45)

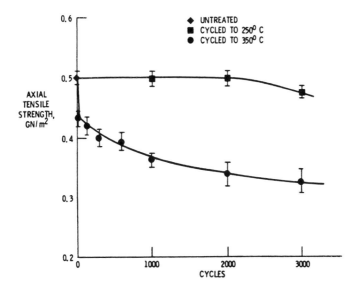

Figure 3.3-4. Room Temperature Axial Tensile Strengths of 55 Volume Percent FP-Al$_2$O$_3$/EZ33Mg Composites after Cycling to 250 °C or 350 °C for Indicated Number of Cycles. Strengths of Untreated Composites are Shown at Zero Cycles (Reference 3-46)

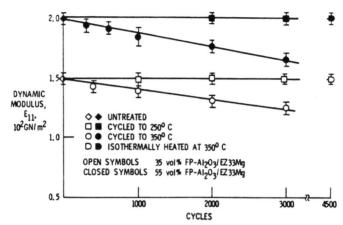

Figure 3.3-5.   Room Temperature Dynamic Moduli of FP-Al$_2$O$_3$/EZ33Mg Composites after Cycling to 250 °C or 350 °C for Indicated Number of Cycles.   Dynamic Moduli of Composites Isothermally Heated at 350 °C for 150 Hours as Shown.   Dynamic Moduli of Untreated Composites are Shown at Zero Cycles (Reference 3-47)

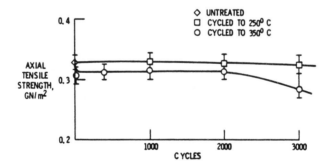

Figure 3.3-6   Room Temperature Axial Tensile Strengths of 35 Volume Percent FP-Al$_2$O$_3$/EZ33Mg Composites after Cycling to 250 °C or 350 °C for Indicated Number of Cycles.   Strengths of Untreated Composites are Shown at Zero Cycles (Reference 3-48)

3.4 Titanium Metal Matrix Composites.

3.4.1 Background. Metals with good ductility and strengths are often combined with fibers of higher strength and stiffness, lower density, and low ductility to provide composites with increased stiffness and lower weight. An excellent example of this combination uses titanium, which has a high specific strength and excellent corrosion resistance, reinforced with boron-based or silicon-carbide-based fibers. The resulting composite possesses significantly increased stiffness and reduced weight. The composite may also have reduced fatigue strength, resulting from many factors including residual stress and the reaction between the fibers and the matrix during the high temperatures of fabrication. Recent developments in boron-based and silicon-carbide-based fibers have significantly reduced this problem of low fatigue strength.

Many advances have been made in titanium matrix composite technology in recent years, but more are still needed. Among these is the need to improve transverse strength through better control of the fiber/reaction zone interfacial bond strength. The major advances need to be in the area of fabrication cost reduction. Methods of continuous fabrication are expected to reduce cost by using coated reinforcements and methods of consolidation. Presently, cost seems to be the major obstacle in the application of titanium-matrix composites to advanced high-performance material systems.

3.4.2 Properties of Titanium Matrix Composites. A number of reinforcements for titanium have been examined in work to date. Table 3.4-1 summarizes comparative data for composites fabricated from these reinforcements. Because of the high reactivity of titanium matrix with the reinforcements, these fibers have been limited to the continuous type as opposed to short fibers or whiskers. At the present time only boron-based and silicon-carbide-based fibers are commercially available and are being strongly considered for use with titanium matrix. Fabrication methods have been developed for Titanium-BORSIC fibers that have given longitudinal strengths of over 200,000 psi with 45 to 50 percent by fiber volume and transverse strength of 65,000 psi. The creep and relaxation properties of

Titanium-Boron composite are also important.   Boron-reinforced titanium possesses good creep strength at 1000 °F.

Silicon Carbide-reinforced Titanium with 22 percent by fiber volume possesses a typical tensile strength of 130,000 psi and an elastic modulus of 30 million psi.   Other unique physical and mechanical properties of boron and silicon carbide reinforced titanium can be found from Figures 3.4-1 through 3.4-12 and from Tables 3.4-2 through 3.4-5.

TABLE 3.4-1.   Summary of Properties of Titanium-Matrix Composites
(Reference 3-49)

Reinforcement	Specific strength		Specific stiffness	Ductility	Stability	Fabricability
	Room temp.	High temp.				
Beryllium	Same as matrix	Same to low	High	Good	Good	Good, may be shaped by forging
Molybdenum	Same as matrix	High	Same as matrix	Moderate	Good	Good
Boron	High	High	High	Low	Good	Difficult
Silicon carbide–boron	High	High	High	Low	Good	Difficult
Silicon carbide	High	High	High	Low	Good	Difficult
Alumina	Same as matrix	Moderate	Moderate	Low	Fair	Very difficult

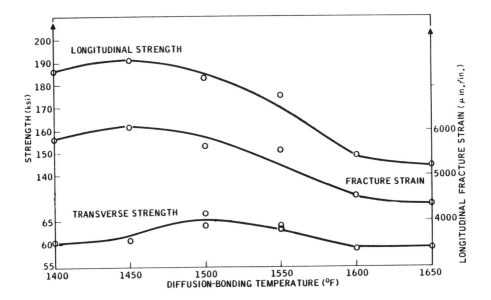

Figure 3.4-1. Effect of Diffusion-Bonding Temperature on Properties of
Ti-6A1-4V-Borsic Composites (Reference 3-50)

TABLE 3.4-2.   Properties of Ti-6Al-4V-50 v/o Borsic Composites
                (Reference 3-51)

Temper-ature (°F)	Orienta-tion (degrees)	Tensile strength (ksi)	Failure strain (μin./in.)	Elastic modulus (10^6 psi)		Coefficient of expansion ($10^{-6}/°$F)
				Tensile	Flexure	
70	0	140	3340	41.5	34.4	2.50
70	15	100	3220	36.8	33.3	—
70	45	66	4220	31.2	31.8	—
70	90	42	3130	29.8	31.2	3.17
500	0	119	—	—	33.2	2.80
700	0	107	—	—	32.4	—
850	0	109	—	—	31.5	3.17
850	15	86	—	—	29.9	—
850	45	53	—	—	27.6	—
850	90	35	—	—	24.4	3.64

TABLE 3.4-3.   Off-Axis Tensile Properties of Ti-6Al-4V-28 v/o SiC
                (Reference 3-52)

Filament orientation (degrees)	Average Strength (ksi)		Elastic modulus (10^6 psi)	Poisson's ratio
	Ultimate tensile strength	Proportional limit		
0	142	117	36	0.275
15	135	117	35	0.277
30	113	104	32	0.346
45	107	75	31	0.346
90	95	53	28	0.250

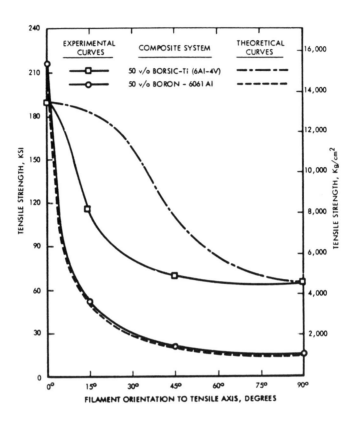

Figure 3.4-2.   Comparison of Off-Axis Tensile Strength of As-Fabricated
B-Al and B/SiC-Ti, using Experimental Data
(Reference 3-53)

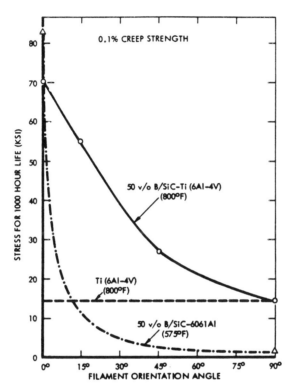

Figure 3.4-3.   Creep-Strength Comparison of Reinforced and Unreinforced
                Titanium at 800 °F (427 °C) and Reinforced Aluminum
                at 575 °F (302 °C) (Reference 3-54)

THE BDM CORPORATION

TABLE 3.4-4.   Tensile (Room-Temperature) Properties of Titanium Matrix
Composites (Reference 3-55)

System	Cycles to Failure, h_f	UTS MPa	ksi
Ti-6Al-4V**	6×10^4	890	129
BORSIC/Ti-6Al-4V	6×10^4	895	130
SiC/Ti-6Al-4V	2×10^4-$2 \times 10^{6\dagger}$	820	119
SCS-6/Ti-6Al-4V	2×10^5	1455	211
B₄C/B/Ti-6Al-4V	1×10^7	1055	153

R = +0.1; σ_{max} = 515 MPa (75 ksi), R.T.

*4-ply, unidirectionally reinforced, 35-40 vol. %
**Mill annealed, 1350°F/2 h/AC
†Significant scatter encountered

TABLE 3.4-5.   Axial Fatigue Properties of As-Fabricated Titanium Matrix
Composites (Reference 3-56)

System	UTS (Long.) MPa	ksi	UTS (Trans.) MPa	ksi	Modulus (Long.)† GPa	msi
Ti-6Al-4V**	890	129	890	129	120	17.5
SiC/Ti-6Al-4V†	820	119	380	55	225	32.6
BORSIC/Ti-6Al-4V†	895	130	365	53	205	30.0
B₄C/B/Ti-6Al-4V†	1055	153	310	45	205	30.0
SCS-6/Ti-6Al-4V†	1455	211	340	49	240	34.8

*4-ply, unidirectionally reinforced, 35-40 vol. %
**Mill annealed, 1350°F/2 h/AC
†After fabrication and low temperature 595°C (1100°F)/512 h exposure

THE BDM CORPORATION

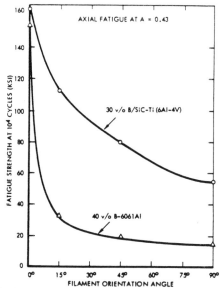

Figure 3.4-4.   Low-Cycle Fatigue Strength of As-Fabricated Aluminum and Titanium Matrix Composites (Reference 3-57)

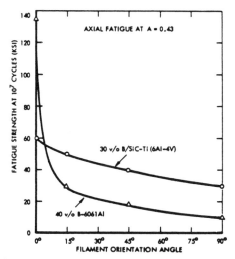

Figure 3.4-5.   High-Cycle Fatigue Strength of As-Fabricated Aluminum and Titanium Matrix Composites (Reference 3-58)

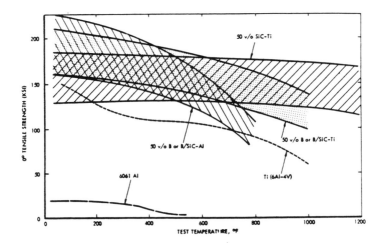

Figure 3.4-6.   Variation of Longitudinal Tensile Strength with Test
Temperature for Unidirectional Aluminum and Titanium
Matrix Composites (Reference 3-59)

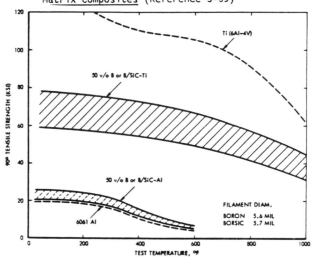

Figure 3.4-7.   Comparison of Transverse Tensile Strength as a Function of
Test Temperature for Aluminum and Titanium Matrix
Composites in As-Fabricated Condition (Reference 3-60)

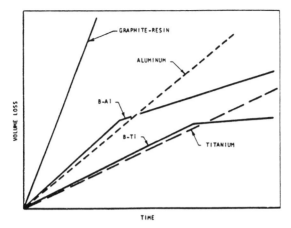

Figure 3.4-8.   Schematic Comparison of Dust Erosion Behavior
                (Reference 3-61)

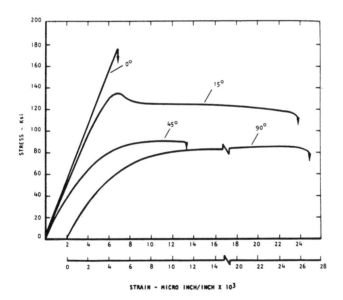

Figure 3.4-9.   Typical Stress-Strain Curves for 30 v/o B/SiC-Ti (6A1-4V)
                Tested at Room Temperature (Reference 3-62)

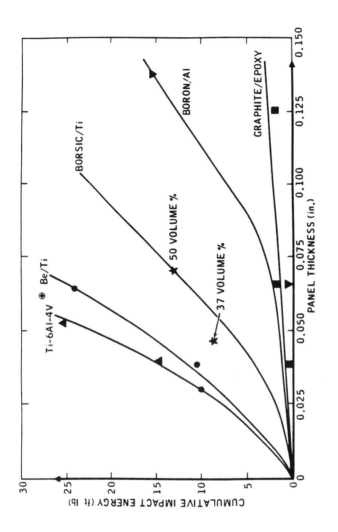

Figure 3.4-10.   Ballistic Impact Strength of Titanium Composites
for ½-in Wide Panels under Static Stress
(Reference 3-63)

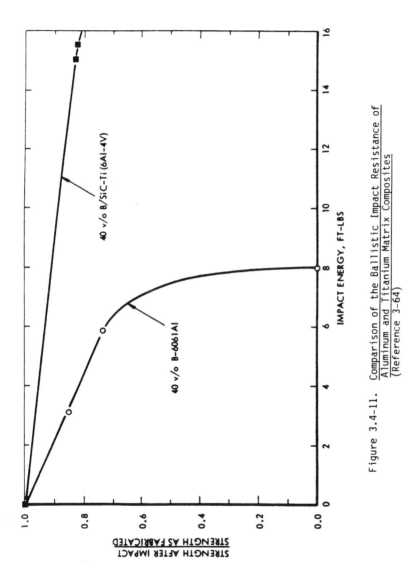

Figure 3.4-11. Comparison of the Ballistic Impact Resistance of Aluminum and Titanium Matrix Composites (Reference 3-64)

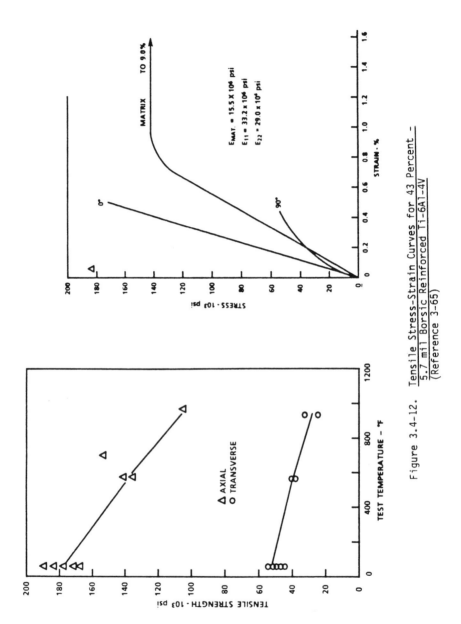

Figure 3.4-12.   Tensile Stress-Strain Curves for 43 Percent –
5.7 mil Borsic Reinforced Ti-6A1-4V
(Reference 3-65)

4. Polymer Matrix Composites

4.1 <u>Introduction</u>. Plastics or polymer matrix composites are quite different from metal matrix composites. Table 4.1-1 compares the key characteristics of polymer matrix composites to metal matrix composites. Plastic or polymer matrix materials result in composites that have higher specific tensile strength and stiffness properties. Plastic composites are more advanced in the fabrication technology and are lower in raw material cost and fabrication cost. There are three important general classes of plastic matrices that are reinforced by the high-performance fiber materials. These are epoxies, phenolics, and polyimides.

Epoxy matrix material can be used with all high-performance fibers. They are cured under a pressure of about 100 psi at a temperature of about 350 °F. Epoxies have been considered quite stable; however, there is some evidence indicating that a loss of properties may occur at temperatures of 250-300 °F as a result of moisture pickup.

Phenolics are highly heat resistant, and composites may be used at temperatures of up to 500 °F for up to 200 hours. Used with glass, high silica, quartz, and graphite fibers, phenolic composites have been employed extensively in nose cones and rocket motor chambers. Difficulty in using reinforcements other than glass occur because of the release of large quantities of volatile substances during curing.

Polyimides are the primary candidates for high-temperature applications. They produce sounder and more heat resistant composites than phenolics. Boron, graphite, and glass fibers have been used with polyimides to provide usable properties at up to 300 °F. Polyimides are powders and must be mixed in a solvent to be impregnated. This presents a problem because the solvent must be removed along with the water to leave a void content of less than three percent. If the process is not carefully controlled, significant strength loss will result. Polyimides are more complex to process than epoxies and require greater pressure and quicker heating.

121

TABLE 4.1-1. Plastic-Matrix Material Characteristics
(Reference 4-1)

Key characteristics of plastic matrix composite (compared to metal matrix composite)	Potential advantage in space applications
Greater strength-to-density and modulus-to-density (due to lower density and high v/o possible)	More efficient in uniaxial tensile loads and bending Less deflection; more resistant to dynamic loads
More advanced in state of the art	More readily available More design data and experience available More test and flight experience available More consistent raw material properties High-strength joints are more fully developed
Lower cost	Cost is approximately one-fourth of that of boron-aluminum
Lower fabrication costs possible	Unit construction possible resulting in less pieces and less joints Less scrap generated (due to cutouts and trimming for metal matrix) More moldable; hence, more shape and thickness variation possible with minimum cost Lower temperature and pressures needed for consolidation
More adaptable to design changes	Changes in loading magnitude or direction can be easily accommodated by adding extra layers onto existing structure
More easily repaired	Less costly to handle. Less impact on schedules if damaged
Matrix is an insulator	Desirable property for thermally-isolated structures, such as spacecraft heat shield substructures

4.2 Epoxy Polymer Matrix Composite.

4.2.1 Background. Epoxy resin developments for composite materials began in the early 1950's. Their excellent adhesion characteristics have resulted in notable success with graphite composites. They also have a good balance of physical and electrical properties and nave a lower order of shrinkage than other thermosetting materials. Added requirements from the aerospace industry for better heat stability and resistance have caused substantial improvements in the materials. All high-performance fibers may be used with epoxy matrices. Epoxies are generally cured at 350 °F, with useful service at up to 350 °F. Minor problems with moisture pickup and loss of some properties may be experienced at 250-350 °F.

4.2.2 Epoxy Resin Preparation. The most common epoxy resin systems are listed below followed by the most frequently used curing agents (Figure 4.2-1). Both of these are important factors for high temperature stability, along with proper curing techniques and procedures.

4.2.2.1 Types of Epoxy Resins.

4.2.2.1.1 Diglycidyl Ethers of Bisphenol (Standard Epoxy). Diglycidyl ethers of bisphenol are the most widely used epoxy resin. While not commonly used for heat resistant forms, widespread use is found in protective coatings, adhesives, sealants, impregnants, bonding, and laminating materials. High molecular weight resins are solid, low molecular weight resins are liquid. Many applications use this resin in conjunction with other types to improve performance.

4.2.2.1.2 Epoxidized Phenolic Novolacs. This resin is available in solid or liquid form, depending on the molecular weight. It generally reacts more quickly than standard epoxies, and it is substantially more heat stable.

4.2.2.1.3 Tetraglycidyl Ether of Tetrakis (Hydroxyphenol) Ether ("Epon" 1031). This epoxy resin brings about better heat stability through greater crosslinking density. The solid "Epon" 1031 is usually used in conjunction with liquid nadic methyl anhydride cure agents to achieve greater fluidity. In the early 1960's, this resin was used to modify standard epoxy resins and was widely used with glass fibers for high temperature epoxy systems.

4.2.2.1.4 Tetraglycidyl Ether of 4,4-Diaminodiphenyl Methane (Ciba's MY-720). This tetrafunctional glycidyl ether is a viscous liquid at room temperature. It has been used in preparing graphite composites and, when reacted with liquid anhydrides, for laminates. It is also used in conjunction with cycloaliphatic epoxies. Water pickup must be avoided during application and lay up because MY-720 and anhydrides are hygroscopic.

4.2.2.2 Curing Agents for Epoxy Resins. A curing agent must be used to considerably decrease curing time. Many different types of curing agents are available, and the final structure may be altered by the choice of curing agents. Curing agents largely determine many behavior characteristics such as chemical resistance and thermal stability. For example, anhydride cures give good electrical insulating properties, thermal stability, and chemical resistance. Aliphatic amines are fast cures and are suited for room temperatures. Aromatic amines give a higher heat resistance, but a higher temperature is also required for cure. Base epoxies with multifunctionality increase the cross-link density and thermal stability. Figures 4.2-2 and 4.2-3 show the importance of using the correct curing agent. Table 4.2-1 compares the properties of different types of cast epoxy resins all cured with aromatic diamines. Other properties of epoxy resins can be found from Figures 4.2-4 through 4.2-7 and in Table 4.2-2.

Structure   Identification

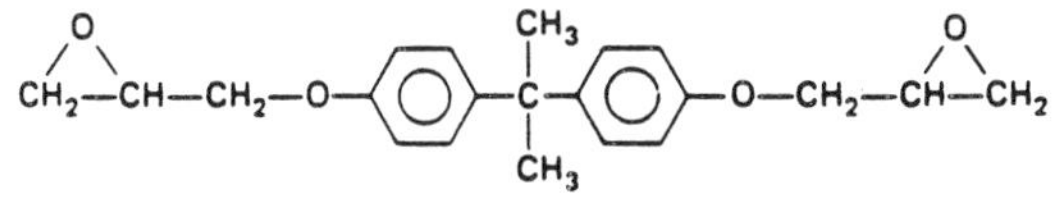

Diglycidyl ether of
bisphenol A
(referred to in this text as
standard epoxy)

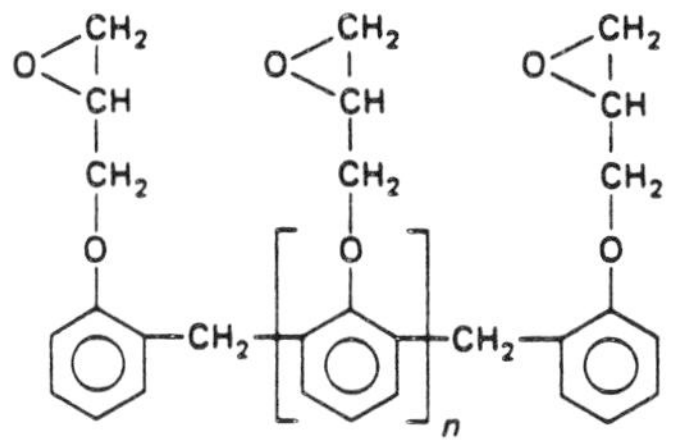

Epoxy novolac
(epoxidized phenolic resin
novolac (example: Dow's
DEN 438)

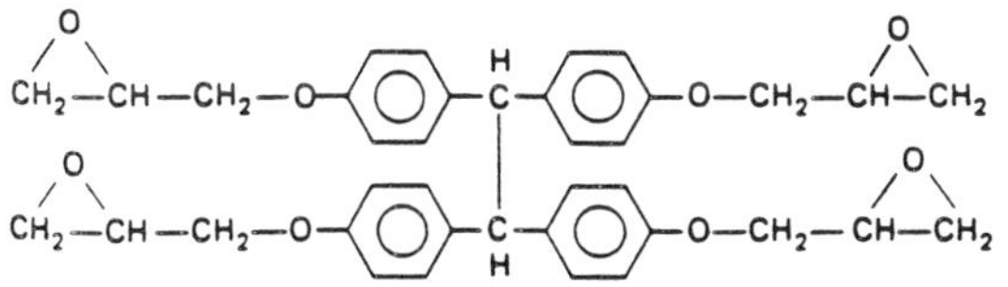

Tetraglycidyl ether of
tetrakis (hydroxy phenyl)
ether
(example: Shell's EPON
1031)

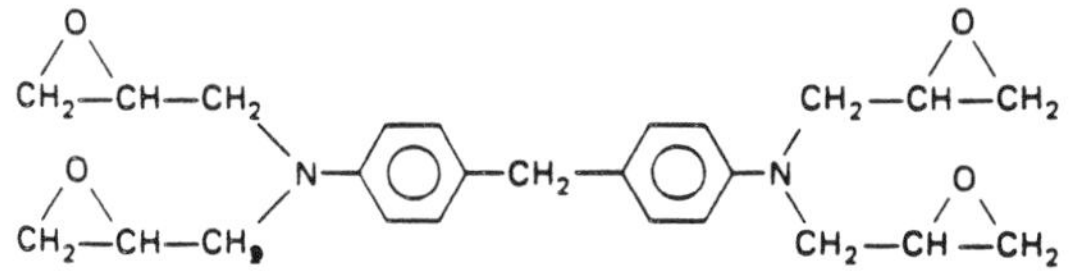

Tetraglycidyl ether of
4,4-diamino diphenyl
methane
(example: Ciba's MY-720)

Figure 4.2-1.   Basic Epoxy Resins Used in Advanced Composites
(Reference 4-2)

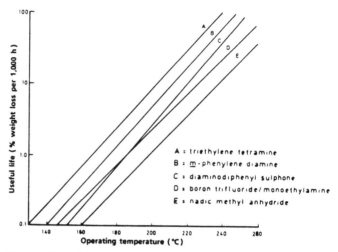

Figure 4.2-2.   Continuous Operating Temperatures for a Bisphenol-A Epoxy Resin
Cured with Different Agents Based on 16 Percent Weight Loss
as End Point   (Reference 4-3)

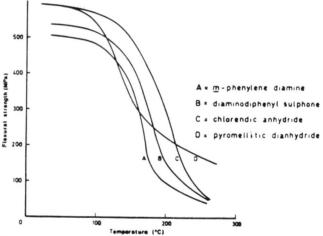

Figure 4.2-3.   Flexural Strength Retention at Temperature of Bisphenol-A
Epoxy Resin/Glass Cloth Laminates Cured with Different
Agents (Reference 4-4)

TABLE 4.2-1. Comparison of Properties of Cast Epoxy Resins (Reference 4-5)

Property	Bisphenol-A epoxy DDM cure	Bisphenol-A epoxy DDS cure	Cyclo aliphatic epoxy MPD cure	Tetrafunctional epoxy DDS cure
Tensile strength (MPa)				
at 20°C	53	59	89	41
150°C	19 (36)	37 (63)	78 (9)	33 (80)
Tensile modulus (MPa)				
at 20°C	2750	3070	6280	4100
150°C	1540 (56)	1470 (48)	432 (7)	3230 (79)
Compressive strength (MPa)				
at 20°C	> 111	107	227	178
150°C	> 29	63 (59)	65 (29)	115 (65)
Compressive modulus (MPa)				
at 20°C	2670	2000	4130	2930
150°C	721 (27)	1280 (64)	1290 (31)	2200 (75)
Flexural strength (MPa)				
at 20°C	116	102	159	86
150°C	41 (35)	49 (48)	84 (53)	86 (100)
Flexural modulus (MPa)				
at 20°C	2730	2790	6450	3920
150°C	1680 (62)	1160 (42)	2640 (41)	3200 (82)
Impact strength (MN/mm)				
at 20°C	0.21	0.17	0.21	0.083
150°C	0.19 (90)	0.21 (100)	0.15 (71)	0.073 (88)
Elongation at break (%)				
at 20°C	4.9	3.3	2.1	1.1
150°C	2.7	8.0	12.3	1.0

Figures in parentheses are percentage retention.
DDM = diaminodiphenylmethane   DDS = diaminodiphenylsulphone
MPD = m-phenylene diamine

TABLE 4.2-2. Key Composite Properties for the Various Epoxy Resins (Reference 4-6)

RESIN MATRIX	4289	828	2256	4617
Cast resin properties				
Initial modulus (10^6 psi)	0.24	0.32	0.60	0.78
Proportional limit (10^3 psi)	1.7	3.2	2.4	7.2
Ultimate elongation (%)	81.0	8.1	6.5	2.2
Ultimate strength (10^3 psi)	5.4	8.0	15.2	14.8
Longitudinal composite properties				
Tensile strength (10^3 psi)	116.0	205.0	207.0	150.0
Tensile modulus (10^6 psi)	13.6	22.4	21.0	19.5
Proportional limit (10^3 psi)	38.1	80.7	70.4	56.7
Compressive strength (10^3 psi)	12.4	134.0	149.0	150.0
Compressive modulus (10^6 psi)	20.0	19.5	19.1	20.9
Proportional limit (10^3 psi)	7.8	44.3	76.2	36.0
Fiber content (% by Vol.)	67.8	62.0	62.7	61.3
Transverse composite properties				
Tensile strength (10^3 psi)	1.2	3.4	3.9	4.3
Tensile modulus (10^6 psi)	0.19	1.18	1.33	1.49
Proportional limit (10^3 psi)	0.08	1.13	2.3	4.0
Torsional stress (10^3 psi)	0.77	3.7	6.9	7.9
Torsional modulus (10^6 psi)	0.054	0.68	0.81	0.96
Proportional limit (10^3 psi)	0.13	1.2	1.7	3.2
Fiber content (% by Vol.)	64.8	68.2	66.7	61.2

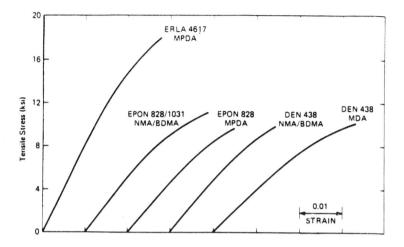

Figure 4.2-4. <u>Tensile Stress-Strain for Epoxy Resins</u> (Reference 4-7)

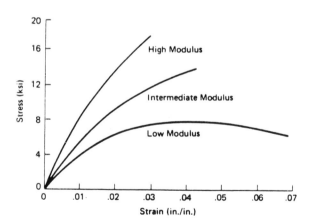

Figure 4.2-5. <u>Stress-Strain Diagrams of Epoxy Matrix Resins</u> (Reference 4-8)

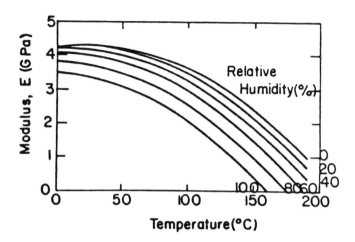

Figure 4.2-6.   Epoxy Matrix Material (Hercules 3501-6) Young's Modulus
                Versus Temperature and Relative Humidity (Reference 4-9)

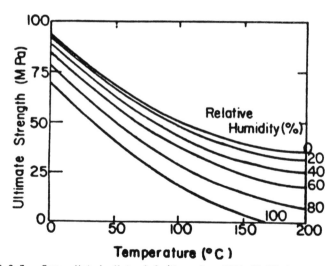

Figure 4.2-7.   Epoxy Matrix Material (Hercules 3501-6) Ultimate Strength
                Versus Temperature and Relative Humidity (Reference 4-10)

4.2.3 Properties of Epoxy Matrix Composites. Epoxy matrix composite materials are recognized for their high performance applications and their compatibility with all fiber types. However, like other composite materials, certain types of epoxies combine best with a specific type of fiber for maximum performance. For example, if the specific requirement is for an increased impact strength or energy absorption composite, a low modulus epoxy would be used. Reasons for the high interest in epoxy matrix composite materials are revealed by the property comparisons with steel, titanium, and aluminum as shown in Figure 4.2-8.

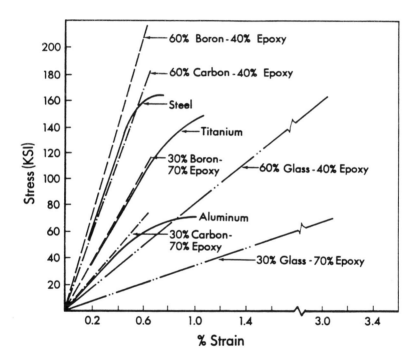

Figure 4.2-8.   Comparison of Epoxy Composite Materials with Steel, Titanium, and Aluminum (Reference 4-11)

4.2.3.1 Graphite-Reinforced Epoxy. Graphite-reinforced epoxy composites provide a combination of strength and stiffness properties that are quite superior to monolithic metallic materials such as aluminum, steel, and titanium. In the graphite/epoxy system, graphite fibers provide the major strength in the direction of the fibers. Several plies of fibers at different angles can be used to provide adequate strength if the structure is designed to carry load along more than one axis. Some typical strength properties can be seen in Figures 4.2-9 and 4.2-10.

The relationship between thermal conductivity and fiber volume fraction for longitudinal and transverse direction of a graphite/epoxy composite is summarized in Figures 4.2-11 and 4.2-12. In general, a high thermal conductivity value is observed along the direction of the fiber, whereas a low conductivity value is observed in the transverse direction.

The electrical characteristics of graphite/epoxy composites are governed by the fiber volume fraction, orientation, void content, and moisture absorption of the epoxy matrix. Examples of typical values of electrical resistivity are given in Figures 4.2-13 and 4.2-14.

The electrical conductivity also governs the electromagnetic shielding characteristics of any material. Shielding effectiveness is defined as a measure of the ability of a material to control the passage of radiated electromagnetic energy. Whenever the composite does not meet the shielding effectiveness design requirements, a more conduction protective coating such as aluminum mesh or aluminum flame spray is used to boost the value of the overall thermal conductivity. An example of this comparison can be found in Figure 4.2-15. Other unique physical and mechanical properties of graphite-reinforced epoxy can be found in Figures 4.2-16 through 4.2-21 and from Tables 4.2-3 through 4.2-5.

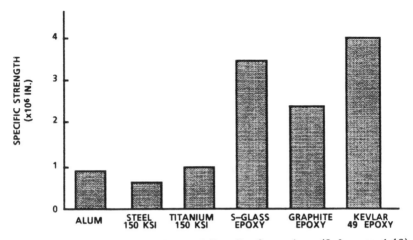

Figure 4.2-9.   Tensile Strength/Density Comparison (Reference 4-12)

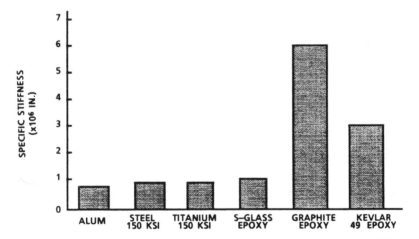

Figure 4.2-10.   Stiffness/Density Comparison (Reference 4-13)

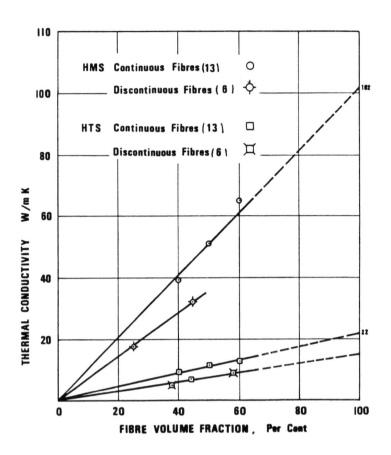

Figure 4.2-11. Longitudinal Thermal Conductivity of Uniaxially Aligned, Continuous and Discontinuous Carbon Fiber-Epoxy Composites at 20 °C as a Function of Fiber Volume Fraction (Reference 4-14)

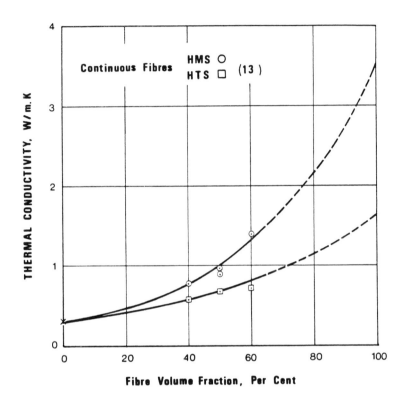

Figure 4.2-12. <u>Transverse Thermal Conductivity of Uniaxially Aligned,</u>
<u>Continuous Carbon Fiber-Epoxy Composites at 20 °C as</u>
<u>a Function of Fiber Volume Fraction</u> (Reference 4-15)

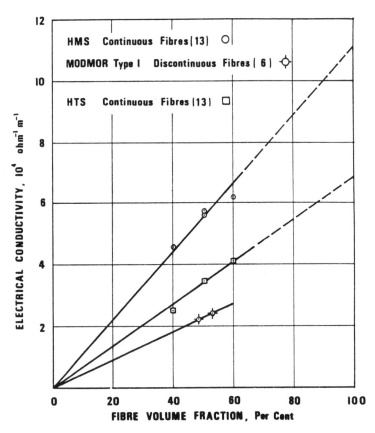

Figure 4.2-13.   Longitudinal Electrical Conductivity of Uniaxially Aligned,
Continuous and Discontinuous Carbon Fiber-Epoxy Composites
at 20 °C as a Function of Fiber Volume Fraction
(Reference 4-16)

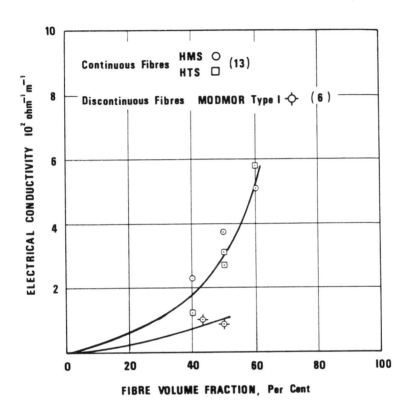

Figure 4.2-14.   Transverse Electrical Conductivity of Uniaxially Aligned,
Continuous and Discontinuous Carbon Fiber-Epoxy Composites
at 20 °C as a Function of Fiber Volume Fraction
(Reference 4-17)

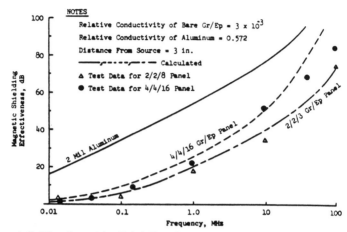

Figure 4.2-15.   Magnetic Shielding Effectiveness for 2/2/8 and 4/4/16
Unprotected Graphite/Epoxy Panels and a 2 Mil Aluminum
Panel (Reference 4-18)

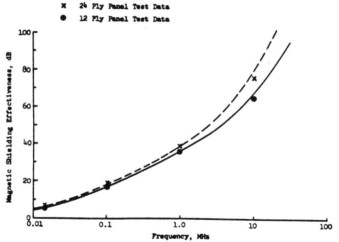

Figure 4.2-16.   Magnetic Shielding Effectiveness for 2/2/8 and 4/4/16
Graphite/Epoxy Panel Protected with 6 Mil Aluminum
Flame Spray (Reference 4-19)

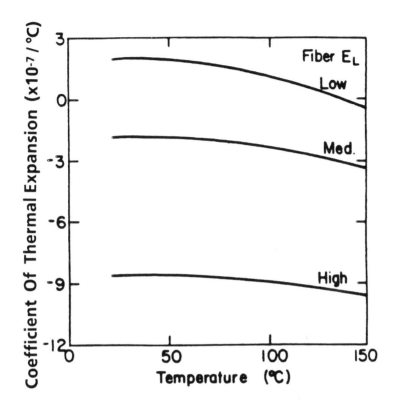

Figure 4.2-17.   Composite Longitudinal Coefficient of Thermal Expansion
Versus Temperature for Three Graphite Fibers
(Reference 4-20)

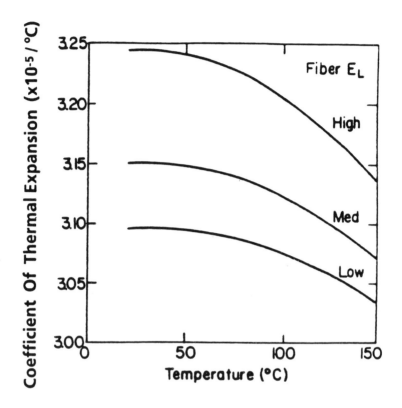

Figure 4.2-18.   Composite Transverse Coefficient of Thermal Expansion
                 Versus Temperature for Three Graphite Fibers
                 (Reference 4-21)

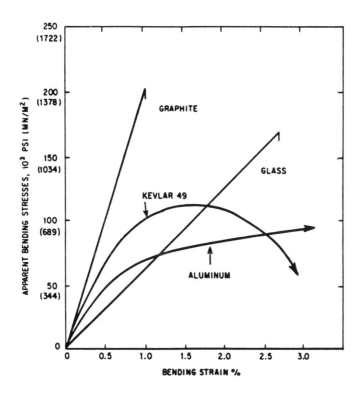

Figure 4.2-19.   Unidirectional Composite Bending Stress/Strain Curves
in Epoxy Resin (Reference 4-22)

TABLE 4.2-3.   Basic Properties of Graphite/Epoxy Composites (0 ± 60°) (Reference 4-23)

			INTERMEDIATE STRENGTH GRAPHITE/EPOXY		HIGH STRENGTH GRAPHITE/EPOXY		HIGH MODULUS GRAPHITE/EPOXY	
			R.T.	350°F	R.T.	350°F	R.T.	350°F
Design strengths (typical)	Longitudinal tensile, ultimate	(ksi)	F_x^{tu}	60.0	45.0	65.0	62.0	42.0	35.0
	Transverse tensile, ultimate	(ksi)	F_y^{tu}	48.0	36.0	52.0	43.0	36.0	24.0
	Longitudinal compression, ultimate	(ksi)	F_x^{cu}	64.0	24.0	69.0	26.0	38.0	22.0
	Transverse compression, ultimate	(ksi)	F_y^{cu}	55.0	21.0	59.0	22.0	33.0	19.0
	In-plane shear, ultimate	(ksi)	F_{xy}^{su}	31.0	23.0	40.0	33.0	26.0	24.0
	Interlaminar shear, ultimate	(ksi)	F^{isu}	–	–	–	–	–	–
	Ultimate longitudinal strain	(μ in./in.)	ϵ_x^{tu}	8,840.0	7,800.0	7,900.0	9,150.0	4,410.0	3,980.0
	Ultimate transverse strain	(μ in./in.)	ϵ_y^{tu}	7,060.0	6,360.0	6,430.0	6,590.0	3,820.0	2,750.0
Elastic properties (typical)	Longitudinal tension modulus	(msi)	E_x^{t}	6.75	5.95	8.09	6.78	9.42	8.73
	Transverse tension modulus	(msi)	E_y^{t}	6.75	5.95	8.09	6.78	9.42	8.73
	Longitudinal compression modulus	(msi)	E_x^{c}	6.75	5.95	8.09	6.78	9.42	8.73

TABLE 4.2-3. (continued)

			INTERMEDIATE STRENGTH GRAPHITE/EPOXY		HIGH STRENGTH GRAPHITE/EPOXY		HIGH MODULUS GRAPHITE/EPOXY	
			R.T.	350°F	R.T.	350°F	R.T.	350°F
Transverse compression modulus	E_y^c	(msi)	6.75	5.95	8.09	6.78	9.42	8.73
In-plane shear modulus	G_{xy}	(msi)	2.58	2.26	3.03	2.57	3.56	3.31
Longitudinal Poisson's ratio	ν_{xy}		0.307	0.318	0.312	0.322	0.324	0.317
Transverse Poisson's ratio	ν_{yx}		0.307	0.318	0.312	0.322	0.324	0.317
Physical constants (typical) Density	ρ	(lb/in.³)	0.055	0.055	0.056	0.056	0.058	0.058
Longitudinal coefficient of thermal expansion	α_x	(μ in./in./°F)	1.79	1.90	1.21	1.10	1.27	0.984
Transverse coefficient of thermal expansion	α_y	(μ in./in./°F)	1.79	1.90	1.21	1.10	1.27	0.984

TABLE 4.2-4. Typical Properties of Various Graphite Epoxy Composites (Reference 4-24)

	Ultimate Properties				
	Tensile		Compression		
	Ultimate, psi	E, psi	Ultimate, psi	E, psi	
Chopped fiber					
molding compound	51 000	15 700 000	68 000	...	
Unidirectional (nonwoven)					
high strength fiber (0°)	236 000	20 000 000	144 000	16 400 000	
high strength fiber (0°/±45°)	72 000	8 300 000	73 000	7 300 000	
medium strength fiber (0°)	211 000	21 300 000	204 000	19 000 000	
medium strength fiber (0°/±45°)	73 000	9 400 000	73 000	9 300 000	
high modulus fiber (0°)	180 000	31 200 000	110 000	25 600 000	
Fabric (woven)					
medium strength fiber	74 000	10 200 000	74 000	9 200 000	

Conversion factor—
10^3 psi = 6.9 MPa, and
10^6 psi = 6.9 GPa.

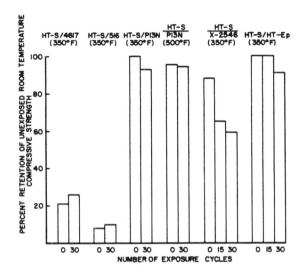

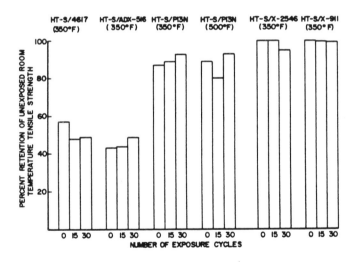

Figure 4.2-20.   Effect of Cyclic Exposures on the Tensile Strengths and Compressive Strengths of Quasi-Isotropic Graphite/Epoxy Composites (Reference 4-25)

TABLE 4.2-5. Typical Impact Energy Values for Graphite/Epoxy and for
Various Materials--Standard Charpy V--Notched Impact Tests
(Reference 4-26)

Material description	Impact energy	
	kJ/m^2	$ft\text{-}lb/in.^2$
Modmor II graphite–epoxy (V_f = 55%)	114	54
Kevlar–epoxy (V_f = 65%)	694	330
S-Glass–epoxy (V_f = 72%)	694	330
Nomex nylon–epoxy (V_f = 70%)	116	55
Boron–epoxy (V_f = 60%)	78	37
4130 Steel alloy (UTS = 100–160 ksi)	593	282
4340 Steel alloy [Rockwell (43–46)]	214	102
431 Stainless steel (annealed)	509	242
2024-T3 Aluminum alloy	84	40
6061-T6 Aluminum alloy (solution treated and precipitation hardened)	153	73
7075-T6 Aluminum alloy (solution treated and precipitation hardened)	67	32

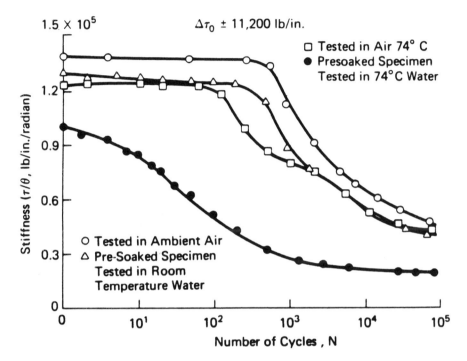

Figure 4.2-21.   Torsion Fatigue of ±45° Fiber-Oriented
                 Graphite-Epoxy Composite. Decrease in
                 Stiffness (τ/θ) with Number of
                 Cycles (log N) as a Function of
                 Environment (Reference 4-27)

4.2.3.2 Kevlar-Reinforced Epoxy. Low density Kevlar fibers used for composite materials provide a typical weight savings of 10 to 30 percent over fiberglass and graphite reinforced materials.   Table 4.2-6 compares epoxy matrix composites reinforced with Kevlar 49, E glass, and graphite. Kevlar 49 shows a greater tensile strength and tensile stiffness, but a lower compressive strength. Composites reinforced with Kevlar 49 also show excellent fatigue behavior, as shown in Figure 4.2-22. Other physical and mechanical properties are shown in Figures 4.2-23 through 4.2-26 and in Tables 4.2-6 through 4.2-8.

TABLE 4.2-6.   Comparison of Kevlar/Epoxy Properties with Different Types
of Composites (Reference 4-28)

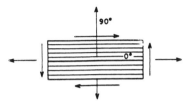

	Kevlar 49	E-Glass	Graphite
Density, lb/in.^{3} (g/cm^3)	0.050 (1.38)	0.075 (2.08)	0.055 (1.52)
Tensile strength 0°, 10^3 psi (MN/m^2)	200 (1 378)	160 (1 102)	180 (1 240)
Compressive strength 0°, 10^3 psi (MN/m^2)	40 (276)	85 (586)	160 (1 102)
Tensile strength 90°, 10^3 psi (MN/m^2)	4.0 (27.6)	5.0 (34.4)	6.0 (41.3)
Compressive strength 90°, 10^3 psi (MN/m^2)	20 (138)	20 (138)	20 (138)
In-plane shear strength 10^3 psi (MN/m^2)	6.4 (44.1)	9.0 (62.0)	9.0 (62.0)
Interlaminar shear strength, 10^3 psi (MN/m^2)	7-10 (48.2-68.9)	12 (82.7)	14 (96.5)
Poisson's ratio	0.34	0.30	0.25
Tensile and compression modulus 0°, 10^6 psi (MN/m^2)	11 (75 790)	5.7 (39 270)	19 (130 900)
Tensile and compression modulus 90°, 10^6 psi (MN/m^2)	0.8 (5 512)	1.3 (8 960)	0.9 (6 200)
In-plane shear modulus, 10^6 psi (MN/m^2)	0.3 (2 070)	0.5 (3 445)	0.7 (4 820)

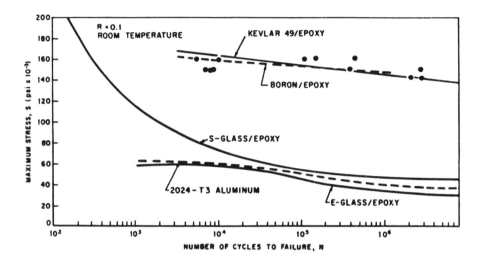

Figure 4.2-22.   Tension-Tension Fatigue Behavior of Unidirectional Kevlar/
Epoxy Composites and Aluminum (Reference 4-29)

TABLE 4.2-7.   Attenuation of Free Vibrations in Kevlar/Epoxy and in
Other Various Materials (Reference 4-30)

Material	Loss Factor $\times 10^4$
1020 steel	>20
Ductile cast iron	30
Graphite/epoxy	30
Fiberglass/epoxy	47
Kevlar 49/epoxy	160
Cured polyester resin	400

$$\text{loss factor} \sim \frac{A_n}{A_{n+1}}$$

Amplitude Decay

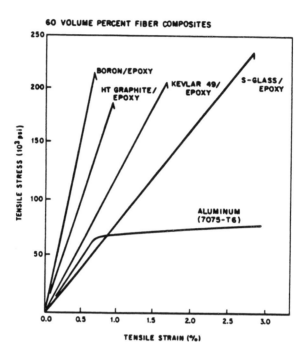

Figure 4.2-23.   <u>Creep and Recovery Behavior of Kevlar 49 Compared to Other Unidirectional Composites</u> (Reference 4-31)

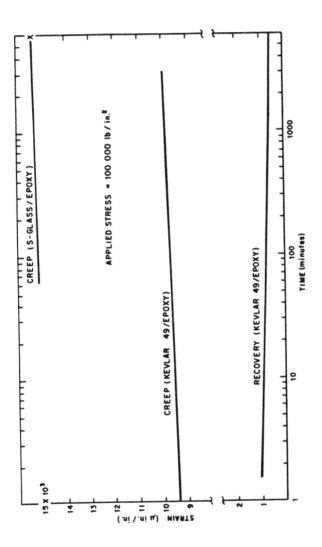

Figure 4.2-24.   Creep and Recovery Behavior of Kevlar 49 and S-Glass Unidirectional Composites (Reference 4-32)

TABLE 4.2-8.   Effect of Form of Reinforcement on Engineering Properties of
Composites of Kevlar 49 Aramid (Reference 4-33)

	Form of Reinforcement of Kevlar 49						
	Unidirectional		Bidirectional		Fabric		Chopped Fiber
Orientation	All 0°		0°/90°		0°/90°		Random
Measurement direction	0°	90°	0°	45°	0°	45°	All
Tensile modulus, 10⁶ psi	11.0	0.8	5.6	0.95	4.5	1.1	3.0
Tensile strength, 10³ psi	200.0	4.0	92.0	14.1	75.0	30.0	28.5
Compressive modulus, 10⁶ psi	11.0	0.8	5.6	1.0	4.5	1.0	...
Compressive strength, 10³ psi	40.0	20.0	29.0	18.3	25.0	18.0	...
Flexural modulus, 10⁶ psi	11.0	...	...	...	4.0	...	2.7
Flexural strength, 10³ psi	90.0	...	...	...	50.0	...	35.5

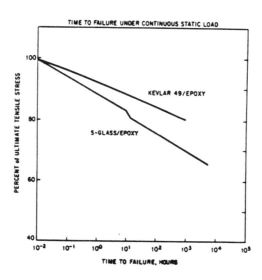

Figure 4.2-25.   Stress-Rupture Behavior of Kevlar 49 and S-Glass
Unidirectional Composites (Reference 4-34)

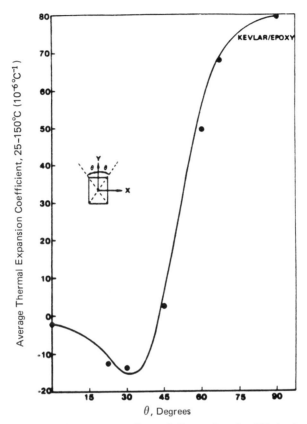

Figure 4.2-26.   The Average Thermal Expansion Coefficient, α_y, Over the Temperature Range 25-150 °C for Bidirectional Kevlar/Epoxy Composites Reinforced in Directions Defined by θ in the Figure. The Solid Curve is a Plot of the Calculated Values of α_y for a Fiber Volume Fraction of 50% (Reference 4-35)

4.2.3.3 Boron-Reinforced Epoxy. Boron-reinforced epoxy systems have exceptional strength at elevated temperatures. They are capable of continuous service at 175 °C (350 °F) and up to 215 °C (420 °F) intermittently.

The boron-epoxy composite material "Narmco 5505" has some interesting results as shown in Table 4.2-9. This material is available as a tape 3 inches (7.62cm) wide and 100 to 250 feet (30 to 76m) long and containing about 620 boron filaments across the width. The thickness of the layers is about .0051 inches (.0130cm). Directional properties make this material an ideal candidate for laminates. Load-carrying capacity and structural weight can be optimized by efficient arrangement and alignment of reinforcing materials.

Comparison of boron-epoxy composites to aluminum alloys, titanium, and fiberglass epoxy composites is shown in Figure 4.2-27. Other unique physical and mechanical properties of boron-reinforced epoxy can be found in Figures 4.2-28 through 4.2-30 and Table 4.2-10.

TABLE 4.2-9. Properties of "Narmco 5505" Boron-Epoxy Composite
(Reference 4-36)

Property	Room temperature		260°F		350°F	
Strength ksi (10^8 N/m²)						
longitudinal tension	198·0	(13·65)	174·0	(12)	149·0	(10·27)
transverse tension	6·5	(0·45)	5·9	(0·4)	4·9	(0·34)
longitudinal compression	230·0	(15·86)	176·0	(12·13)	159·0	(10·96)
transverse compression	30·9	(2·13)	19·4	(1·38)	14·5	(1·0)
in-plane shear	9·0	(0·62)	5·0	(0·35)	3·0	(0·21)
Modulus 10^6 psi (10^{10} N/m²)						
longitudinal tension	30·6	(21·1)	30·3	(20·9)	25·6	(17·65)
transverse tension	3·5	(2·41)	2·1	(1·45)	1·0	(0·69)
longitudinal compression	34·0	(23·4)	33·0	(22·75)	32·0	(22·06)
transverse compression	3·7	(2·55)	2·35	(1·62)	1·5	(1·0)
in-plane shear	1·00	(0·69)	0·80	(0·55)	0·22	(0·15)
major Poisson's ratio	0·36		0·35		0·30	
minor Poisson's ratio	0·033		0·025		0·017	
Coefficient of thermal expansion 10^{-6} in./in.°F (10^{-6} cm/cm°C)						
longitudinal	2·5	(4·5)	2·5	(4·5)	2·5	(4·5)
transverse	13·1	(23·6)	17·7	(31·9)	20·2	(36·4)

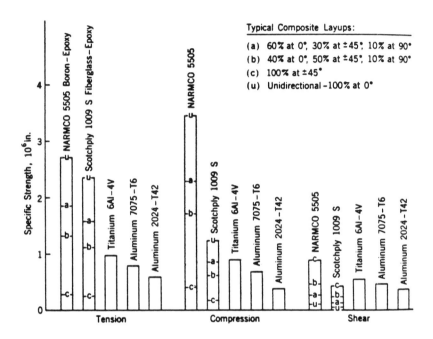

Figure 4.2-27.   Specific Strength of Boron/Epoxy Composite Materials and Metals at Room Temperature   (Reference 4-37)

TABLE 4.2-10. Allowable Strengths of a Unidirectional Boron Epoxy
Laminate (Reference 4-38)

	Room Temperature	350°F
0° tensile strength (10^3 psi)	230	180
0° tensile modulus (10^6 psi)	32	28
90° tensile strength (10^3 psi)	10	5
90° tensile modulus (10^6 psi)	3.0	1.4
0° compression strength (10^3 psi)	360	110
Interlaminar shear (10^3 psi)	16	8

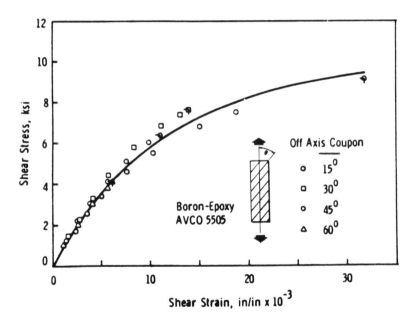

Figure 4.2-28. Boron/Epoxy Composite Shear Stress-Strain
(Reference 4-39)

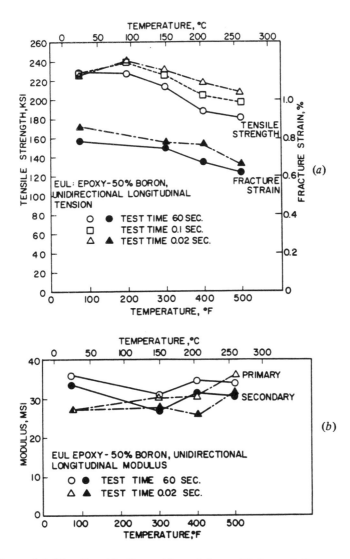

Figure 4.2-29. Tensile Properties of Epoxy-50 Percent Boron Unidirectional Composite at Several Temperatures and Strain Rates. (a) Longitudinal Ultimate Strength and Fracture Strain; (b) Longitudinal Young's Moduli (Reference 4-40)

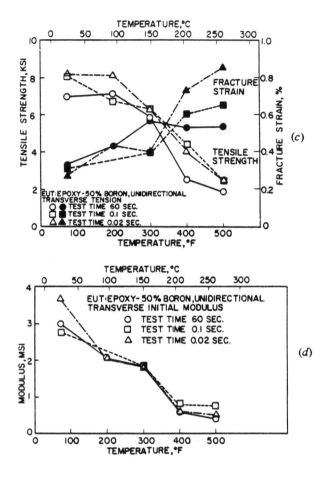

Figure 4.2-29. (continued) (c) Transverse Ultimate Strength and Fracture Strain; (d) Transverse Young's Moduli

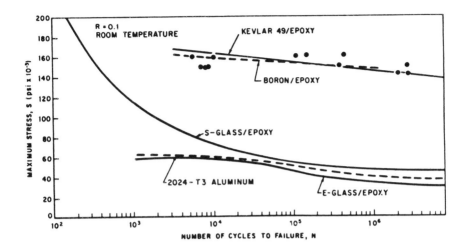

Figure 4.2-30.   Tension-Tension Fatigue Behavior of Unidirectional
Boron/Epoxy Composite and Aluminum (Reference 4-41)

4.2.3.4 Glass-Reinforced Epoxy. In the aerospace industry, more
advanced composite parts are manufactured with fiberglass reinforcement
than with any type of advanced fiber. Fiberglass reinforced composites are
of great interest to structural designers because of the fiber long-term
aging and strength retention properties, high tensile and impact strengths
composites, and high chemical resistance. Table 4.2-11 shows the fiberglass and quartz-reinforced composite properties with 50 percent fiber volume fraction. The disadvantages of using fiberglass are low modulus, self-
abrasiveness of fibers, relatively low fatigue resistance, and poor
adhesion to matrix resins.

TABLE 4.2-11.   Typical Properties of Glass/Epoxy and Quartz/Epoxy
Composites (Reference 4-42)

Type of Fiber	E-Glass	S-Glass	Quartz
Density, g/cm^3	0.065	0.060	0.065
Tensile strength, unidirectional, psi	160 000	200 000	. . .
Tensile strength, fabric, psi	60 000	85 000	70 000
Tensile modulus, unidirectional, psi	6 000 000	8 000 000	
Tensile modulus, fabric, psi	3 400 000	4 000 000	3 000 000
Compressive strength, fabric, psi	50 000	60 000	40 000
Dielectric constant	4.3	3.8	3.2
Loss factor	0.19 to 0.24	0.25	0.01
Coefficient of thermal expansion			
in./in./°F	4.8×10^6	3.5×10^6	2.2×10^6
cm/cm/°C	8.6×10^6	6.3×10^6	4.0×10^6

Conversion factor—
10^3 psi = 6.9 MPa, and
in./in./°F = 0.556 cm/cm/°C.

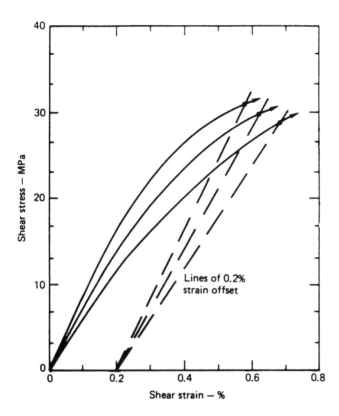

Figure 4.2-31. Average ±45 °Tensile Shear Stress-Strain
Curves for Three-Volume Percentages of S-2 Glass Fiber
in a Room-Temperature-Curable Epoxy Matrix (Reference
4-43)

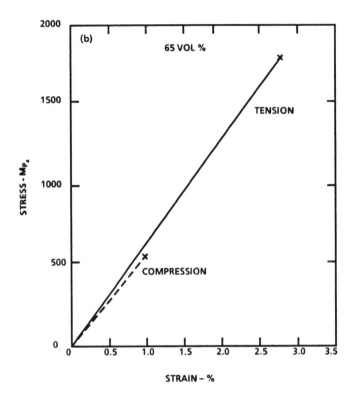

Figure 4.2-32.   Average Longitudinal Tensile and Compressive Curves for 65% Fiber Volume (Reference 4-44)

TABLE 4.2-12.   Comparison of Mechanical Properties for Five S-2 Glass-Epoxy Composites (Reference 4-45)

MECHANICAL PROPERTY	COMPOSITE (FIBER CONTENT)				
	50 VOL%	55 VOL%	60 VOL%	65 VOL%	70 VOL%
ELASTIC CONSTANTS					
LONGITUDINAL MODULUS, GPa	45.23 ± 0.89[a] (31)	49.75 ± 0.98[a] (31)	54.28 ± 1.05[a] (31)	58.80 ± 1.16[a] (31)	63.32 ± 1.25[a] (31)
TRANSVERSE MODULUS, GPa	12.73 ± 0.24[b] (34)	14.28 ± 0.24[b] (34)	15.85 ± 0.24[b] (34)	17.46 ± 0.24[b] (34)	19.09 ± 0.24[b] (34)
SHEAR MODULUS, GPa	~4.18[c] (20)	~5.11[c] (20)	6.14 ± 0.19[c] (20)	7.28 ± 0.19[c] (20)	8.52 ± 0.19[c] (20)
MAJOR POISSON'S RATIO	0.2653 ± 0.0057[a] (31)	0.2651 ± 0.0057[a] (31)	0.2653 ± 0.0057[a] (31)	0.2651 ± 0.0057[a] (31)	0.2655 ± 0.0057[a] (31)
MINOR POISSON'S RATIO	0.0749 ± 0.0020[b] (34)	0.0763 ± 0.0021[b] (34)	0.0777 ± 0.0022[b] (34)	0.0789 ± 0.0023[b] (34)	0.0802 ± 0.0024[b] (34)
ULTIMATES					
LONGITUDINAL TENSION					
STRESS, MPa	1346 ± 52[a] (29)	1480 ± 57[a] (29)	1615 ± 62[a] (29)	1749 ± 67[a] (29)	1884 ± 72[a] (29)
STRAIN, %	2.93 ± 0.14 (6)	3.07 ± 0.02 (6)	3.12 ± 0.09 (9)	2.87 ± 0.27 (10)	2.57 ± 0.32 (5)
LONGITUDINAL COMPRESSION					
STRESS, MPa	380 ± 50[d] (4)	420 ± 50[d] (4)	460 ± 60[d] (4)	500 ± 60[d] (4)	540 ± 70[d] (4)
STRAIN, %	–	–	–	0.93 ± 0.07 (4)	–
TRANSVERSE TENSION					
STRESS, MPa	35.62 ± 0.95[b] (29)	37.76 ± 0.95[b] (29)	39.82 ± 0.95[b] (29)	41.81 ± 0.95[b] (29)	43.75 ± 0.95[b] (29)
STRAIN, %	0.314 ± 0.010[b] (29)	0.294 ± 0.010[b] (29)	0.277 ± 0.010[b] (29)	0.263 ± 0.010[b] (29)	0.250 ± 0.010[b] (29)
TRANSVERSE COMPRESSION					
STRESS, MPa	–	–	–	111.6 ± 2.3 (5)	–
STRAIN, %	–	–	–	3.89 ± 0.47 (5)	–
± 45°-TENSILE SHEAR, AT 0.2% OFFSET					
STRESS, MPa	~26.7[c] (20)	~27.9[c] (20)	29.04 ± 0.39[c] (20)	30.12 ± 0.39[c] (20)	31.16 ± 0.39[c] (20)
STRAIN, %	~0.82[c] (20)	~0.74[c] (20)	0.676 ± 0.014[c] (20)	0.620 ± 0.014[c] (20)	0.573 ± 0.014[c] (20)

[a] BY RULE OF MIXTURES FROM SPECIMENS RANGING FROM 41 TO 65 VOL% FIBER.

[b] INTERPOLATED FROM TESTS OF SPECIMENS RANGING FROM 44 TO 68 VOL% FIBER.

[c] EXTRAPOLATED OR INTERPOLATED FROM TESTS OF SPECIMENS RANGING FROM 62 TO 69 VOL% FIBER.

[d] BY RULE OF MIXTURES FROM RESULTS AT 63.4 VOL% FIBER.

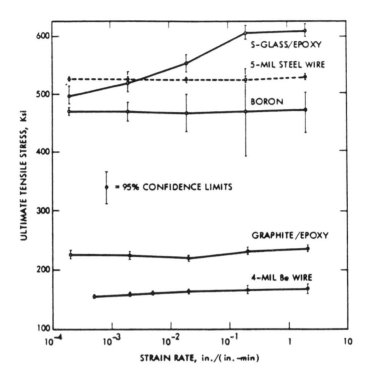

Figure 4.2-33. <u>Strain Rate Effect on the Ultimate Tensile Stress
of Glass Fiber/Epoxy Strands</u> (Reference 4-46)

TABLE 4.2-13. Comparison of Thermal Property Data for Three S-2 Glass/
Epoxy Composites (Reference 4-47)

THERMAL PROPERTY	COMPOSITE (FIBER CONTENT)		
	60 VOL %	65 VOL%	70 VOL
LINEAR COEFFICIENT OF THERMAL EXPANSION, 10^{-6}/°C LONGITUDINAL (-50 to +75°C)	3.523 ± 0.038	3.523 ± 0.038	3.379 ± 0.38
TRANSVERSE -50°C -25°C 0°C 25°C 50°C 75°C	23.5 ± 0.8 24.3 ± 0.8 26.3 ± 0.8 28.9 ± 0.8 32.7 ± 3.4 98 ± 21	19.9 ± 0.8 21.3 ± 0.8 23.2 ± 0.8 25.1 ± 0.8 27.6 ± 3.4 77 ± 21	21.3 ± 0.8 23.4 ± 0.8 24.9 ± 0.8 27.6 ± 0.8 29.6 ± 3.4 86 ± 21
THERMAL CONDUCTIVITY, W/m °C LONGITUDINAL 0°C 25°C 50°C 75°C	1.50 ± 0.26 1.58 ± 0.26 1.67 ± 0.26 1.75 ± 0.26	1.62 ± 0.26 1.70 ± 0.26 1.80 ± 0.26 1.88 ± 0.26	1.74 ± 0.26 1.82 ± 0.26 1.92 ± 0.26 2.00 ± 0.26
TRANSVERSE -50°C -25°C 0°C 25°C 50°C 75°C	0.477 0.509 0.540 0.571 0.603 0.634	- - - - - -	- - - - - -

4.3 Phenolic Polymer Matrix Composite.

4.3.1 Background. Although phenol formaldehyde resins (phenolic resins) have not found as widespread use as epoxy resins, they are still used
in many applications. Epoxy has advantages of better adhesion, lower cure
shrinkage, and the absence of voids; but these advantages may be overlooked
for the unique char strength of phenolic matrix composites. The physical
integrity of the charred structure makes phenolics desirable for such
applications as ablative shields for reentry vehicles and rocket nozzles.
Phenolic resins may be used as binders for carbon fiber structures. They
may also be used as curing agents for some epoxies because of their

hydroxyl group.   The rate of reaction of epoxy resin polymerization is accelerated by the presence of phenol or thio-phenol.   Epoxidized phenolic-novolacs are candidates for advanced composite matrix material.   Flammability and smoke evolution have recently been minimized in phenolics, making them excellent candidates for applications where these properties are important.

    4.3.2 Phenolic Resin Preparation. Compared with other thermosetting materials, phenolic resins are very inexpensive.   The production of phenolic resins involves a condensation reaction between phenol or phenol mixture and an aldehyde, such as formaldehyde, in the presence of a catalyst. The ratio of the reactants and the type of catalyst can be manipulated to form either a linear or a network structure.   The structure is usually in the form of phenol molecules linked together by methylene and ether bridges.   Many production methods make use of a partially cured powdered resin that allows fusion and cross-linking to form a final continuous structure by the application of heat and pressure.   Impact strength and shrinkage are controlled by the addition of fillers.   An idealized phenolic structure is shown below (Figure 4.3-1) where n = 0-2, w = 0-3, x = 0-2, y = 0-1.   Other unique properties of phenolic resin can be found in Table 4.3-1 and Figure 4.3-2.

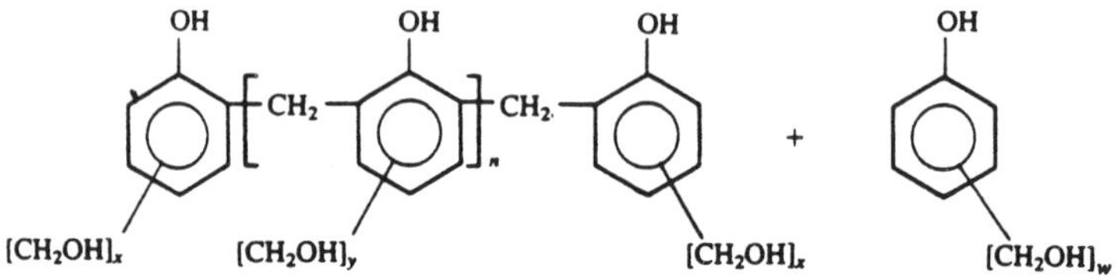

Figure 4.3-1. Formation of Phenolic Resin (Reference 4-48)

TABLE 4.3-1.   Thermal Conductivity of Phenolic and Typical
Thermosetting Resins at 35 °C (Reference 4-49)

Material	Density $g\,cm^{-3}$	Thermal conductivity $W\,m^{-1}\,K^{-1}$
Phenolic	1·36	0·27
	1·25	0·29
Epoxy	1·22	0·20
	1·18	0·29
	1·22	0·26
Polyester	1·21	0·18

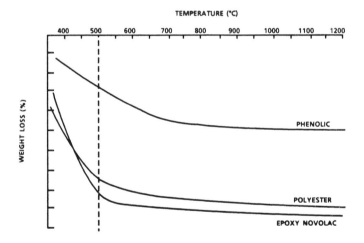

Figure 4.3-2.   Comparison of Thermal Stabilities of Polyester,
Phenolic, and Epoxy Novolac Resins.   (Based on weight
loss after 30 minutes at temperature up to 500 °C and
5 minutes at temperature thereafter) (Reference 4-50)

Phenolic fibers are available commercially under the trade name Kynol. These fibers will not melt or burn, and when exposed to flame, they give off 60 percent carbon residue with major volatiles of carbon dioxide and water (Figure 4.3-3). Blends with other fibers are often used to decrease flammability. Weight loss of Kynol fibers at 150, 200, and 250 °C is shown in Figure 4.3-4. It should be noted that temperatures above 150° cause oxidative degradation, and 150 °C should be the maximum temperature for long term use. Addition of compounds preventing peroxide formation will cause some improvement. Fiber properties after exposure to high temperatures are given in Table 4.3-2. In inert atmospheres, the tensile strength and modulus are not significantly changed by long exposures at high temperatures, but elongation at break is substantially reduced.

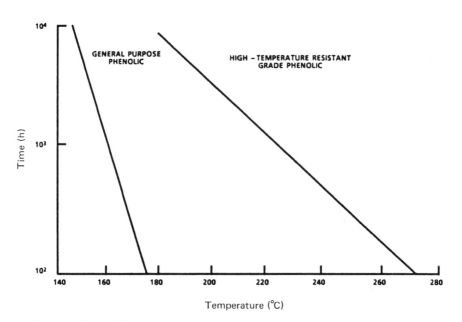

Figure 4.3-3.   Lifetime at Temperature Plots for Phenolic Moldings Based Upon 70 Percent Retention of Flexural Strength (Reference 4-51)

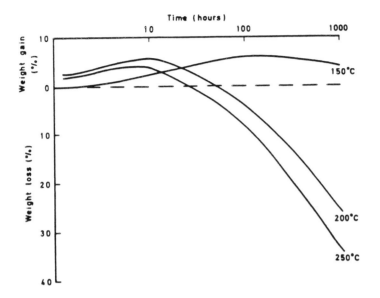

Figure 4.3-4.   Isothermal Weight Loss of Phenolic Kynol Fibers in Air
                at Different Temperatures (Reference 4-52)

TABLE 4.3-2.   Properties of Phenolic Kynol Fibers after Aging at
               Elevated Temperatures (Reference 4-53)

Aging condition	Atmosphere	Tenacity (g/den)	Modulus (g/den)	Elongation at break (%)
Unaged		1.9	48	33.6
350 hours at 250°C	Nitrogen	1.9	44	7.3
500 hours at 250°C	Nitrogen	1.9	42	9.2
1000 hours at 250°C	Nitrogen	1.9	46	8.3
100 hours at 350°C	Nitrogen	1.8	43	6.8
500 hours at 150°C	Air	0.9	47	4.2
1200 hours at 150°C	Air	0.6	37	1.6
350 hours at 250°C	Air	0.7	35	2.9
500 hours at 250°C	Air	0.7	49	2.6
1000 hours at 250°C	Air	0.4	30	1.2

4.3.3 Properties of Phenolic Matrix Composites. The aerospace industry has extensively used phenolic matrix composites. They are primarily
reinforced with glass fabric and have been successfully used at temperatures of 500 °F for up to 200 hours. Other reinforcements have been used
in thermal protection applications including nylon, high-silica, quartz,
and graphite textile reinforcements.

Phenolics release large amounts of volatiles during the condensation
reaction, thus presenting much difficulty in the use of reinforcements
other than glass. Phenolic composites are not as sound as polyimides and
do not have as great heat resistance. Thermal degradation of phenolics can
be characterized by either the amount of volatiles given off or by weight
loss. Figure 4.3-5 describes the weight loss of a glass-phenolic system at
various high temperatures as a function of time. Decomposition is accelerated uniformly with increasing temperatures above 300 °F. Figure 4.3-6
shows mechanical properties of carbon reinforced phenolics as a function of
temperature. Flexural strength tends to have an accelerated decrease above
120 °C, but torsional modulus changes very little, even at 240 °C. Various
properties of glass fiber reinforced phenolic matrix composite are listed
in Table 4.3-3. With the exception of flexural strength and modulus, there
is relatively little change in properties. In fact, tensile and compressive rupture strengths are retained very well at 260 °C for over 1000
hours. This is superior to aluminum alloys commonly used for structural
materials in aircraft. Figure 4.3-7 shows changes in flexural, compressive, tensile strength, and modulus of silane-modified phenolic/glass fiber
composite relative to temperatures up to 535 °C. Seen from this represen-
tation, phenolic resins seem to have excellent retention of properties at
high temperatures, but a difference is significant when held at these
temperatures for extended periods of time. Other unique properties of
phenolic matrix composites can be found in Table 4.3-4 through 4.3-8 and
Figures 4.3-8 and 4.3-9.

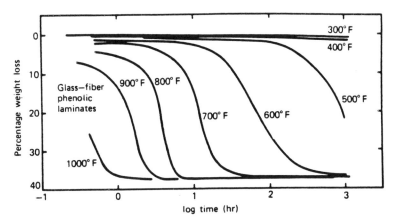

Figure 4.3-5.   Thermal Degradation of Glass-Phenolic System
(Reference 4-54)

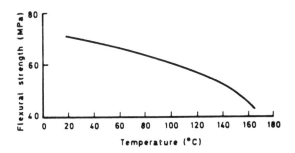

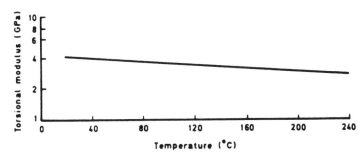

Figure 4.3-6.   Mechanical Properties as a Function of Temperature of
Carbon-Filled Phenolic Moldings (Reference 4-55)

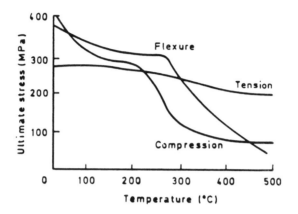

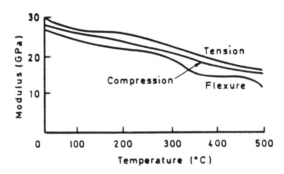

Figure 4.3-7.   Change in Mechanical Properties of a Silane-Modified
Phenolic Resin/Glass Fiber (181 Volan A) Laminate with
Temperature.   Tested at Temperature After 30 Minutes at
Temperature (Reference 4-56)

TABLE 4.3-3. Properties of a CTL 91-LD Phenolic Resin/181 Glass Fiber
Laminate (Reference 4-57)

Property	Room temperature	Value at 260°C after 30 minutes at 260°C	Property retention (%)
Flexural strength (MPa)	600	449	75
Flexural modulus (GPa)	33	26	79
Tensile strength (MPa)	400	359	90
Tensile modulus (GPa)	23	21	91
Compressive strength (MPa)	345	319	92
Shear strength (MPa)	286	266	93
Izod impact strength (J/mm notch)	0.8	0.8	100
Dielectric constant at 1 MHz	4.0	3.98	—
at 10 MHz	3.57	4.16	—
Power factor at 1 MHz	0.0098	0.0055	—
at 10 MHz	0.010	0.0126	—

TABLE 4.3-4. Transverse and Longitudinal Thermal Conductivity of a
Glass Fiber Reinforced Phenolic Resin ($V_F = 0.54 \pm 0.02$) (Reference 4-58)

Temperature K	Thermal conductivity $W m^{-1} K^{-1}$		k_{CL}/k_{CT}	Density $g\ cm^{-3}$
	k_{CT}	k_{CL}		
300	0.508	0.60	1.18	2.18
250	0.472	0.548	1.16	—
200	0.428	0.490	1.14	—
150	0.374	0.423	1.13	—
100	0.320	0.357	1.12	—
50	0.27	0.285	1.06	—

TABLE 4.3-5. Vibration Data for Carbon-Phenolic (warp direction)
(Reference 4-59)

Mode n	Frequency f_n (Hz)	Modulus $E_1 \times 10^{-6}$ (psi)	Loss Factor tan δ
1	77	3.18	0.0054
2	480	3.21	0.0083
3	1,335	3.23	0.0077
4	2,585	3.24	0.0095
5	4,230	3.19	0.0107
6	6,230	3.17	0.0123
7	8,600	3.25	0.0140
8	11,200	3.22	0.0150

TABLE 4.3-6. Vibration Data for Silica-Phenolic (warp direction)
(Reference 4-60)

Mode n	Frequency f_n (Hz)	Modulus $E_1 \times 10^{-6}$ (psi)	Loss Factor tan δ
1	118	3.32	0.0051
2	741	3.35	0.0079
3	2060	3.33	0.0092
4	3983	3.35	0.0124
5	6426	3.30	0.0146
6	9426	3.32	0.0179

TABLE 4.3-7.   High-Temperature Flexural Properties of Carbon-Based
Plastic Laminates (Reference 4-61)

MATERIAL	DENSITY (lb/ft³)	FLEXURAL STRENGTH (lb/in²)				FLEXURAL MODULUS (10⁶lb/in²)			
		75°F	400°F	700°F		75°F	400°F	700°F
Phenolic[a] 40% Carbon fabric[b] 60%	89.8	26,500	26,500	21,200		2.27	1.49	1.31
Phenolic[a] 35% Carbon fabric[c] 65%	82.0	18,300	17,500	13,600		0.92	1.06	0.82
Phenyl silane[d] 32% Carbon fabric[c] 68%	83.5	18,600	19,600	14,600		1.19	1.10	0.89
Phenolic[a] 30% Graphite fabric[e] 70%	82.6	20,200	17,600	8160		1.64	1.40	0.65
Phenyl silane[d] 30% Graphite fabric[e] 70%	83.2	19,900	17,200	10,600		1.58	1.51	1.04

[a]Monsanto "SC 1008."
[b]"CCA-1."
[c]"VCA."
[d]C.T.L. "37-9X."
[e]"WCB."

TABLE 4.3-8.   Strength Properties of Carbon-Based Fabric Reinforced Plastics (Reference 4-62)

MATERIAL	DENSITY (lb/ft^3)	TENSILE STRENGTH (lb/in^2)	FLEXURAL STRENGTH (lb/in^2)	FLEXURAL MODULUS (10^6lb/in^2)	COMPRESSIVE STRENGTH (lb/in^2)	COMPRESSIVE MODULUS (10^6lb/in^2)
Phenolic[a] 40% Carbon fabric[b] 60%	89.8	16,600	26,500	2.27	40,200	2.47
Phenolic[a] 35% Carbon fabric[c] 65%	82.0	9160	18,300	0.92	23,800	1.50
Phenolic[a] 46% Carbon fabric[c] 54%	85.2	7250	14,800	1.01	27,100	1.22
Epoxy novolac[d] 27% Carbon fabric[c] 73%	78.7	9310	16,000	0.93	16,700	1.19
Phenolic[e] 27% Graphite fabric[f] 73%	82.6	11,600	20,100	1.69	13,200	2.13
Phenolic[e] 37% Graphite fabric[f] 63%	86.7	9970	17,900	1.17	10,900	1.75
Phenyl silane[g] 30% Graphite fabric[f] 70%	83.2	11,200	19,900	1.58	11,500	1.95

[a]Monsanto "SC 1008."
[b]"CCA-1."
[c]"VCA."
[d]Dow Chemical "D.E.N. 438."
[e]C.T.L. "91LD."
[f]"WCB."
[g]C.T.L. "37-9X."

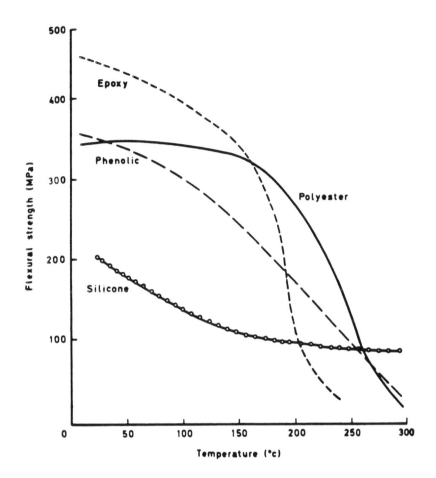

Figure 4.3-8. Flexural Strength of Various Glass-Cloth Laminates after 100
Hours at Temperature. (All heat-cleaned Y227 cloth.
Polyester resin-PDL-7-669. Epoxy resin-Epikote 828.
Phenolic resin-V17085. Silicone resin MS 2106.)
(Reference 4-63)

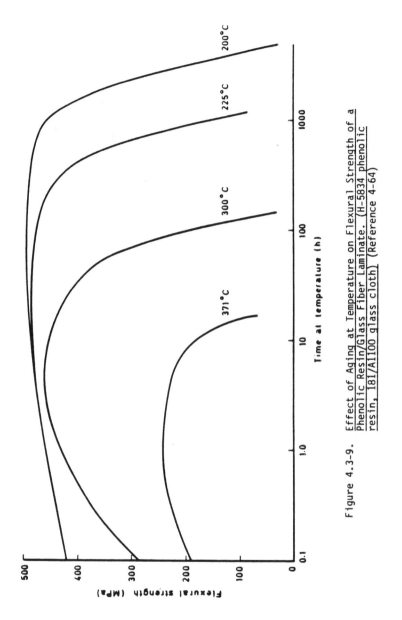

Figure 4.3-9.   Effect of Aging at Temperature on Flexural Strength of a
Phenolic Resin/Glass Fiber Laminate. (H-5834 phenolic
resin, 181/A1100 glass cloth) (Reference 4-64)

4.4 Polyimide Polymer Matrix Composites.

4.4.1 Background. Polyimide resins are candidate materials for use with graphite reinforcements to form composite systems capable of use at 600 °F (316 °C). Polyimides cost more than epoxies, but this difference can be overlooked because of other more significant properties.

Polyimides are synthesized by a reaction between an aromatic tetracarboxylic dianhydride, such as PMDA or BTDA, and an aromatic diamine. This reaction forms a polyamic acid, which is soluble in dimethyl acetimide or a similar solvent. The prepolymer is cured at a high temperature, forming an aromatic linear resin (Figure 4.4-1). Polyimides are currently the prime candidate for high temperature advanced resin composite applications.

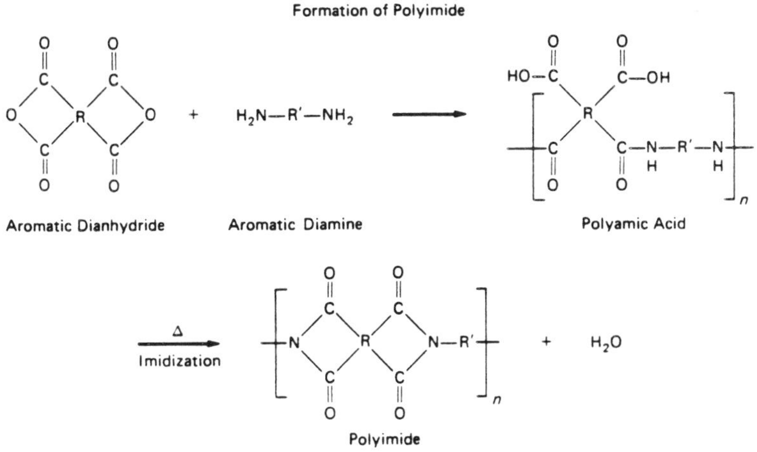

Formation of Polyimide

Figure 4.4-1. Formation of Polyimide (Reference 4-65)

4.4.2 Polyimide Resin Preparation. Early work in polyimides focused on the production of condensation type polyimides. These are produced by the reaction of an aryldianhydride with an aryldiamine, producing a polyamide acid through chain extension. Polymerization takes a long time, and the products are unstable, even in solution, and must be refrigerated.

More recently, partial esters of polyanhydrides have been used, eliminating the evolution of water and other volatiles. This process involves a higher curing temperature and a higher thermal stability which are due to reactive end groups capable of cross-linking. Thermoplastic polyimides are classified as linear polymers with thermoplastic processing qualities. They do not need a post-cure cycle, but do require high press temperatures and possibly high pressures. Vacuum bag/autoclave processing has not been developed for all of these. systems. Some mechanical properties of polyimide resins are given in Tables 4.4-1 and 4.4-2.

TABLE 4.4-1.   Comparison of Unreinforced Polyimide Matrix Properties
(Reference 4-66)

Property	Temperature (°C)	Condensation	Thermoplastic Type 1 (NR150)	Thermoplastic Type 2 (2080)
Tensile strength (MPa)	25	76–100	110	118
	260	—	—	30
	288	—	—	28
	300	33–43	—	—.
	316	—	31	—
Tensile modulus (GPa)	25	3.10–3.35	4.00–4.66	1.30
	260	—	—	—
	288	—	—	0.67
	300	1.90–2.07	—	—
	316	—	1.04	—
Compressive strength (MPa)	25	253–310	—	206
	300	128–133	—	—
Compressive modulus (GPa)	25	3.93	—	2.04
	300	1.90	—	—
Flexural strength (MPa)	25	107–128	117	119
	200	—	—	124
	250	—	—	55
	288	—	—	35
	300	58–72	—	—
Flexural modulus (GPa)	25	3.03–3.17	3.80–4.17	3.32
	200		2.42	2.00
	250		1.93	1.59
	288		1.66	1.11
	300	1.79–1.89	—	—
Impact strength (J/cm) Izod notched		0.38–0.76	0.43	0.38
Elongation at break (%)	25	5–8	6	10
	300	3	[316°C] 65	
Heat distortion temperature (°C)		357	—	270–280
Water pick-up (% in 24 hours)		0.32	—	0.6

TABLE 4.4-2.   Comparison of Unreinforced Polyimide Properties (Reference 4-67)

Property	Temperature (°C)	Addition		
		Norbornene/PMR	Bismaleimide	Acetylenic
Tensile strength (MPa)	25	48–83	48–59	83
	260	39	—	—
Tensile modulus (GPa)	25	3.80	—	3.93
	260	2.39	—	—
Compressive strength (MPa)	25	255	197	173
Compressive modulus(GPa)	25	2.89	—	—
Flexural strength (MPa)	25	76	128	131
	200	—	83	—
	250	48	57	—
	260	41–55	—	—
	288	35–41	—	—
	316	—	—	29
Flexural modulus (GPa)	25	3.17–3.38	3.80	4.49
	200	—	3.17	—
	250	2.17	2.76	—
	260	2.07–2.28	—	—
	288	1.93–2.07	—	—
Elongation at break (%)	25	1.4–2.5	<1	2
	260	1.8	—	—
Heat distortion temperature (°C)		>300	—	195–210
Water pick-up (% in 24 hours)		0.4	—	[1000 hours] < 1

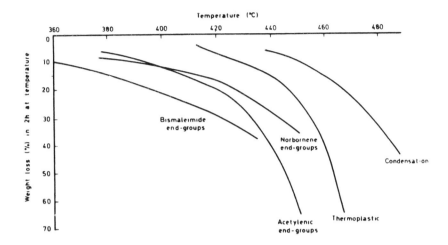

Figure 4.4-2. Comparison of Thermal Stability in Air of Different Types of Polyimide (Reference 4-68)

4.4.3 Properties of Polyimide Matrix Composites. PMR polyimides were the first high-temperature polymer matrix systems that met with any degree of success.   These polymer matrix composites provided heat resistance nearly double that of existing epoxy systems. Unfilled polyimides have low wear resistance, but with 10 percent graphite, wear resistance becomes good and is retained well in high temperatures. Mechanical properties of different polyimides and carbon reinforcements are given in Table 4.4-3. Differences in resin properties do not significantly affect properties in their composites. Table 4.4-4 shows interlaminar shear strength as a function of time and temperature. As technology improves, minor advances in these areas may occur, but major developments are not expected. The development of boron-reinforced polyimides has proved difficult because the resin tends to flow out during curing. Processing polyimides is in general more complex than epoxy resins. Careful steps must be followed to keep void content low by eliminating volatiles and avoiding loss of resin. Other unique properties of polyimide matrix composites can be found in Figure 4.4-2 through 4.4-5 and Tables 4.4-5 and 4.4-6.

TABLE 4.4-3. Comparison of Properties of Carbon Fiber/Polyimide Resin Unidirectional Laminates (Reference 4-69)

Polyimide type	Carbon fiber type	Interlaminar shear strength (MPa)	Flexural strength (MPa)	Flexural modulus (GPa)
Condensation	High modulus	49	710	163
	High strength	72	1340	115
	Type III or A	66	1420	108
Addition—norbornene	High modulus	52	—	—
	High strength	78	1430	100
Addition—bismaleimide	High strength	85	1410	120
Addition—acetylenic	High strength	83	1350	104
	Type III or A	116	1190	108
Thermoplastic type I	High modulus	51	870	145
(NR 150)	High strength	61	1240	108
	Type III or A	88	1390	104

TABLE 4.4-4. Currently Achievable Levels of Performance with Polyimide/Carbon Fiber Composites (Reference 4-70)

Polyimide type	Interlaminar shear strength (MPa)	Temperature (°C)	Time (h)
Condensation	83	260	1000
Addition–PMR	83	260	1000
Addition–bismaleimide	83	230	1000
Condensation	55	350	10
Condensation	55	320	1000
Thermoplastic	55	320	1000
Addition–PMR	55	320	500
Addition–PMR	55	290	1000
Thermoplastic	55	260	50000
Addition–PMR	55	260	10000
Additon–bismaleimide	55	260	1000
Thermoplastic	28	350	10
Addition–PMR	28	350	10
Addition–PMR	28	320	1000
Condensation	28	290	10000
Thermoplastic	28	290	10000
Addition–PMR	28	290	10000
Thermoplastic	28	230	50000
Addition–PMR	28	230	50000
Addition–bismaleimide	28	230	10000

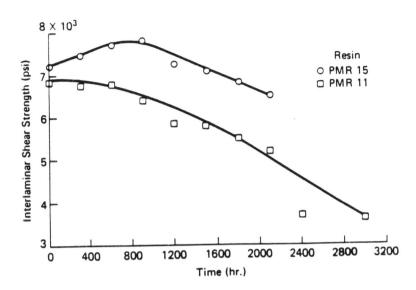

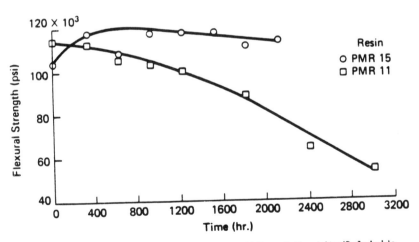

Figure 4.4-3.   Flexural Strength Stability of Graphite/Polyimide
Composites at 600 °F (Reference 4-71)

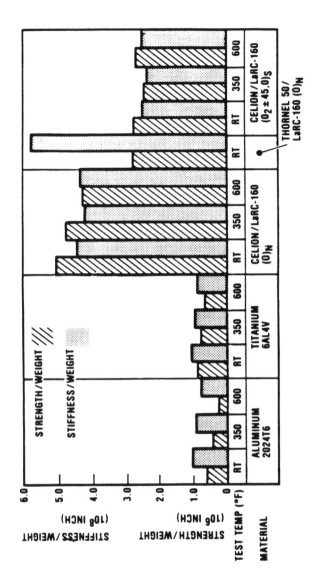

Figure 4.4-4. Graphite/Polyimide Composite System Capable of 600 °F Service Temperature Could Replace Baseline Aluminum Structural Material. (Reference 4-72)

TABLE 4.4-5.   Graphite/Polyimide Composite Properties (Reference 4-73)

Composite System		Manufacturer	T_g, °C (°F)	ΔT, °C (°F)	Fiber Content, vol %	Void Content, vol %
Resin	Fiber	Prepreg/Composite				
NR-150B2	△ C-6000	Fiberite Corp./United Technologies Corp. (Hamilton Standard Div.)	337 (638)	291 to 343 (556 to 649)	60.1 ± 2.2	3.1 ± 0.5
	▲ GY-70 (i)	Fiberite Corp./United Technologies Corp. (Hamilton Standard Div.)	365 (689)	322 to 386 (612 to 727)	61.8 ± 1.4	3.9 ± 0.2
	▼ GY-70 (ii)	Composites Horizons/Composites Horizons	346 (654)	294 to 361 (561 to 682)	55.0 ± 0.2	~0
PMR-15	○ C-6000	Composites Horizons/Composites Horizons	380 (717)	269 to 405 (516 to 761)	59.2 ± 1.4	4.5 ± 0.8
	⊙ F-5A	Composites Horizons/Composites Horizons	379 (714)	279 to 401 (534 to 754)	55.8 ± 2.0	3.7 ± 1.0
	● GY-70	Composites Horizons/Composites Horizons	371 (700)	294 to 397 (561 to 747)	56.1 ± 0.3	3.7 ± 0.4
CPI-2237	□ C-6000	Ferro Corp./Ferro Corp.	376 (708)	283 to 385 (542 to 725)	63.7 ± 1.3	4.8 ± 0.6
	■ GY-70	Ferro Corp./Ferro Corp.	379 (715)	296 to 386 (565 to 727)	57.3 ± 1.0	1.9 ± 0.6

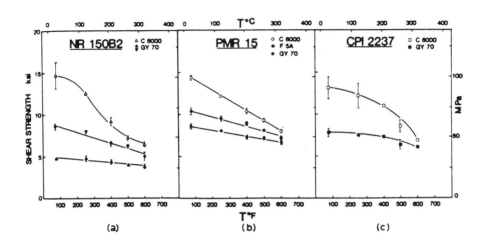

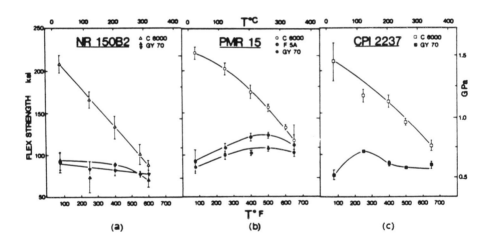

Figure 4.4-5.   Strength Dependence on Temperature of Graphite/Polyimide
Composites, Shown According to Matrix Resin Type
(Reference 4-74)

TABLE 4.4-6.   Test Matrix, Test Panel Properties and Test Results for Different Types of Graphite/Polyimide Matrix Composites (Reference 4-75)

Layup	Thermal Conductivity			Thermal Expansion			Specific Heat	Emittance
	X	Y	Z	X	Y	Z	...	...
UNI	1	1	...	2	2	...	...	...
ISO	1	...	1	2	...	2	2	2

NOTE—A similar set was tested for both HTS/PMR 15 and HTS/NR 150B2. Numbers indicate specimen quantities. Temperature extremes are: 116 and 588 K.

Material	Layup	Total Plies	Nominal Thickness, mm	Resin Content, wt %	Tensile Strength, MN/m^2
HTS/PMR 15	(UNI) $[0]_{18}$	18	3.2	35.1	1 263.0
HTS/PMR 15	(ISO) $[0,45,90,135]_{2B}$	16	2.8	32.9	460.0
HTS/NR 150B2	(UNI) $[0]_{24}$	24	3.2	22.2	1 551.0
HTS/NR 150B2	(ISO) $[0,45,90,135]_{B2}$	24	3.2	23.1	548.0

Material	Layup/Direction	Thermal Conductivity, w/mk			Thermal Expansion[a] 10^{-6} m/m/K	Specific Heat		
		170 K	300 K	540 K		180 K	300 K	590 K
HTS/PMR 15	UNI X	13	22	42	−0.2/0	...	...	...
		...	...	...	−0.4/0.6	...	...	...
HTS/NR 150B2	UNI X	4.6	7.6	14	−0.5/0.5	...	...	...
		...	...	...	−0.4/0.2	...	...	...
HTS/PMR 15	UNI Y	0.85	0.93	1.4	24.0	...	...	...
		...	...	...	24.0	...	...	...
HTS/NR 150B2	UNI Y	0.60	0.86	1.3	16.0	...	...	...
		...	...	...	16.0	...	...	...
HTS/PMR 15	ISO X	6.6	9.9	15	0.8/1.7	...	...	...
		...	...	...	0.8/1.5	0.440	0.808	1.36
HTS/NR 150B2	ISO X	2.2	4.8	9.6	0.3/1.0	...	...	...
		...	...	...	0.4/1.2	0.419	0.775	1.24
HTS/PMR 15	ISO Z	0.33	0.57	0.85	30.0	...	...	...
		...	...	...	−4/20	...	...	...
HTS/NR 150B2	ISO Z	0.33	0.55	0.87	17.0	...	...	...
		...	...	...	23.0	...	...	...

[a] Where there were significant differences between slopes above and below ambient temperature two values are listed.

References

Chapter 2

2-1.   Broutman, L., and Krock, R., (eds.) <u>Modern Composite Materials</u>, Addison Wesley Publisher, New York, NY, 1969.

2-2.   Agarwal, B., and Broutman, L., <u>Analysis and Performance of Fiber Composites</u>, John Wiley and Sons, New York, NY, 1980.

2-3.   Delmonte, J., <u>Technology of Carbon and Graphite Fiber Composites</u>, Van Nostrand Reinhold Co., New York, NY, 1981.

2-4.   Watts, A. (ed.) <u>Commercial Opportunities for Advanced Composites</u>, STP 704, ASTM Special Technical Publication, 1980.

2-5.   Watts, A., p. 6.

2-6.   Watts, A., p. 9.

2-7.   Watts, A., p. 9.

2-8.   Donnet, J., and Bansal, R., <u>Carbon Fibers</u>, Marcel Dekker Inc., New York, NY, 1984, p. 214.

2-9.   Donnet, J., p. 316

2-10.   Zweben, C., <u>Metal Matrix Composites</u>, Part 1, UCLA Extension Short Course Program, June 11-15, 1984.

2-11.   Donnet, J., p. 201.

2-12.   Watts, A., p. 11.

2-13.   Lynch, C., and Kershaw, J., <u>Metal Matrix Composites</u>, CRC Press, Cleveland, OH, 1972.

2-14.   Cook, J., and Sakurai, T., 10th SAMPE National Symposium, 1966.

2-15.   Zweben, C., p. 216.

2-16.   Watts, A., p. 23.

2-17.   Watts, A., p. 26.

2-18.   Watts, A., p. 28.

2-19.   Watts, A., p. 29.

2-20.   Watts, A., p. 29.

2-21.   Watts, A., p. 14.

2-22.   Watts, A., p. 14.

2-23.   Watts, A., p. 16.

2-24.   Watts, A., p. 19.

2-25.   Otto, W., "Properties of Glass Fibers at Elevated Temperatures," Proceedings of the 6th Sagamore Ord. Materials Research Conference, 1959.

Chapter 3

3-1.   Korb, L., "Space Vehicles." Composite Materials, Vol. 3, Academic Press, New York, NY, 1974.

3-2.   Watts, A., Commercial Opportunities for Advanced Composites, STP 704, ASTM Special Technical Publication, 1980.

3-3.   Schoutens, J., Discontinuous Silicon Carbide Reinforced Aluminum Metal Matrix Composites Review, MMCIAC No. 000461, December 1984.

3-4.   Schoutens, J., p. 3-2.

3-5.   Renton, W., "Hybrid and Select Metal Matrix Composite Fabrication: State-of-the-Art Review." AIAA Publication, 1977.

3-6.   Renton, W., p. 283.

3-7.   Kendall, E., "Development of Metal Matrix Composites Reinforced with High Modulus Graphite Fibers." Composite Materials, Vol. 4, Academic Press, New York, NY, 1974.

3-8.   Renton, W., p. 315.

3-9.   Metcalfe, A., "Fiber Reinforced Titanium Alloys." Composite Materials, Vol. 4, Academic Press, New York, NY, 1974.

3-10.   Metcalfe, A., p. 18.

3-11.   Zweben, C., Metal Matrix Composites, Part I, UCLA Extension Program, June 11-15, 1984.

3-12.   Zweben, C., p. 152.

3-13.   Toth, I., "Comparison of the Mechanical Behavior of Filamentary Reinforced Aluminum and Titanium Alloys." STP 546, ASTM Special Publications, 1973.

3-14.   Zweben, C., p. 163.

3-15.   Watts, A., p. 80.

3-16.   Watts, A., p. 81.

3-17.   Meyn, D., "Effect of Temperature and Strain Rate on the Tensile Properties of Boron/Aluminum and Boron/Epoxy Composites." STP 546, ASTM Special Technical Publication, 1973.

3-18.   Kreider, K., Composite Materials, Academic Press, New York, NY, 1974.

3-19.   Piggott, M., Load Bearing Fiber Composites. Pergamon Press, Oxford, 1980.

3-20.   Zweben, C., p. 168.

3-21.   Zweben, C., p. 192.

3-22.   Kreider, K., p. 450.

3-23.   Kreider, K., p. 455.

3-24.   Hilado, C., Boron Reinforced Aluminum Systems, Materials Technology Series, Vol. 6, 1974.

3-25.   Hilado, C., p. 440.

3-26.   Hilado, C., p. 442.

3-27.   Rack, J., "Discontinuous Metal Matrix Composite Fabrication." UCLA Metal Composite Short Course, 1983.

3-28.   Rack, J., p. 52.

3-29.   Nieh, T., "Creep Rupture of a Silicon Carbide Reinforced Aluminum Composite." Metallurgical Transactions, No. 15A, 1984.

3-30.   Nieh, T., p. 139.

3-31.   Schoutens, J., p. 5-85.

3-32.   Schoutens, J., p. 5-84.

3-33.   Leonard, B., "Metal Matrix Composites for Undersea Systems." Metal Matrix Composites (4th Conference), Arlington, VA, 1981.

3-34.   Gubbay, J., _Micromechanical Properties of Al/SIC Metal Matrix Composites_, CSDL-R-1578 Final Report, MA, 1982.

3-35.   Schoutens, J., p. 7-32.

3-36.   Foltz, J., _Metal Matrix Composites for Naval Systems_, NSWC-MP-79-15, Silver Spring, MD, 1978.

3-37.   Gubbay, J., p. 315.

3-38.   Nowitzky, A., _Discontinuous SIC Whisker Aluminum Composite Material and Structural Shapes: Fabrication and Characterization_, AFWAL-TR-81-3182, 1982.

3-39.   Schoutens, J., p. 5-45.

3-40.   Huseby, I., Shyne J., "Technical Note on Some Mechanical Properties of Magnesium-Boron particulate composite" _Journal of Powder Metallurgy_, Vol. 9, No. 2, 1973, p. 91.

3-41.   Coutts, W., "Material Property Design Data for FP/Magnesium Composites," Tech. Report No. AMMRC TR 82-54, 1982, p. 33.

3-42.   Attmad, I., Barranco, J., "Reinforcement of Magnesium with Boron and Titanium Filaments." _Journal Metal Trans._, Vol. 4, March 1973, p. 793.

3-43.   Ahmad, I., p. 796.

3-44.   Coutts, W., p. 43.

3-45.   Coutts, W., p. 42.

3-46.   Bhatts, R., "Thermal Degradation of the Tensile Properties of Unidirectionally Reinforced $FP-Al_2O_3/EZ$ 33 Magnesium Composites." _Mechanical Behavior of Metal Matrix Composites_, AIME Publication, 1983.

3-47.   Bhatts, R., p. 56.

3-48.   Bhatts, R., p. 56.

3-49.   Metcalfe, A., p. 313.

3-50.   Metcalfe, A., p. 305.

3-51.   Metcalfe, A., p. 304.

3-52.   Metcalfe, A., p. 300.

3-53.   Toth, I., p. 549.

3-54.   Toth, I., p. 559.

3-55.   Smith, P., "Developments in Titanium Metal Matrix Composites," Journal of Metals, March, 1984.

3-56.   Smith, P., p. 23.

3-57.   Toth, I., p. 555.

3-58.   Toth, I., p. 555.

3-59.   Toth, I., p. 552.

3-60.   Toth, I., p. 552.

3-61.   Toth, I., p. 553.

3-62.   Toth, I., p. 550.

3-63.   Metcalfe, A., p. 281.

3-64.   Toth, I., p. 554.

3-65.   Jaffee, R., Burte, H., Titanium Science and Technology, Volume 4, Plenum Press, New York, 1973.

Chapter 4

4-1.   Korb, L., "Space Vehicles," Composite Materials, Volume 4, Academic Press, New York, NY, 1974.

4-2.   Delmonte, J., Technology of Carbon and Graphite Fiber Composites, Van Nostrand Reinhold Co., New York, NY, 1981.

4-3.   Brenner, W., Lum, D., and Riley, M., High Temperature Plastics, Reinhold Publishing Company, New York, NY, 1962.

4-4.   Lee, H., and Neville, K., Handbook of Epoxy Resins, McGraw-Hill, New York, NY, 1967.

4-5.   Berry, D., Buck, B., and Cornwell, A., Handbook of Resin Properties, Part A: Cast Resins, Yarsley Testing Lab, Ashstead, 1975.

4-6.   Delmonte, J., p. 106.

4-7.   Delmonte, J., p. 108.

4-8.   Delmonte, J., p. 313.

4-9.   Adam, D. and Monib, M., "Moisture Expansion and Thermal Expansion
        Coefficients of a Polymer-Matrix Composite Material, Fibrous
        Composites in Structural Design, 1980.

4-10.   Adam, D., p. 28.

4-11.   McCullough, R., Concepts of Fiber-Resin Composites, Marcel
        Dekker, Inc., New York, NY, 1971.

4-12.   Richardson, M., Polymer Engineering Composites, Applied Science
        Publishers LTD, London, Eng., 1977.

4-13.   Richardson, M., p. 89.

4-14.   Richardson, M., p. 347.

4-15.   Richardson, M., p. 348.

4-16.   Richardson, M., p. 351.

4-17.   Richardson, M., p. 354.

4-18.   Volpe, V., "Estimation of Electrical Conductivity and Electromag-
        netic Shielding Characteristics of Graphite/Epoxy Laminate,
        Materials Technology Series, Vol. 9, 1982.

4-19.   Volpe, V., p. 106.

4-20.   Adam, D., p. 18.

4-21.   Adam, D., p. 16.

4-22.   Watts, A. (ed.), Commercial Opportunities for Advanced Compos-
        ites, STP 704, ASTM Special Technical Publication, 1980.

4-23.   Delmonte, J., p. 218.

4-24.   Watts, A., p. 47.

4-25.   Browning, C., and Hartness, J., "Effects of Moisture on the Prop-
        erties of High Performance Structural Resins and Compos-
        ites," Composite Materials: Testing and Design (3rd Confer-
        ence), STP-546, ASTM Special Technical Publications, 1973.

4-26.   Agarwal, B., and Broutman, L., Analysis and Performance of Fiber
        Composites, John Wiley and Sons, New York, NY, 1980.

4-27.   Delmonte, J., p. 294.

4-28.   Watts, A., p. 58.

4-29.   Watts, A., p. 63.

4-30.   Watts, A., p. 64.

4-31.   Watts, A., p. 59.

4-32.   Watts, A., p. 61.

4-33.   Watts, A., p. 46.

4-34.   Watts, A., p. 62.

4-35.   Strife, J., and Prewo, K., "The Thermal Expansion Behavior of Unidirectional and Bidirectional Kevlar/Epoxy Composites," Carbon Reinforced Epoxy Systems Part 3, 1982.

4-36.   Hadcock, R., "Boron/Epoxy Aircraft Structures," Handbook of Fiberglass and Advanced Plastics Composites, Lubin, G. (ed.), Van Nostrand Reinhold Co., New York, NY, 1969.

4-37.   Hadcock, R., p. 73.

4-38.   Watts, A., p. 51.

4-39.   Pipes, R., "Interlaminar Fatigue Characteristics of Fiber-Reinforced Composite Materials," Composite Materials: Testing and Design (3rd Conference), ASTM STP 546, 1974.

4-40.   Meyn, D., "Effect of Temperature and Strain Rate on the Tensile Properties of Boron/Aluminum and Boron/Epoxy Composites," Composite Materials: Testing and Design (3rd Conference), ASTM STP 546, 1974.

4-41.   Watts, A., p. 63.

4-42.   Watts, A., p. 53.

4-43.   Clements, L., and Moore, R., "Composite Properties for S-2 Glass at Room Temperature-Curable Epoxy Matrix," SAMPE Quarterly, 1979.

4-44.   Clements, L., p. 30.

4-45.   Clements, L., p. 32.

4-46.   Hilado, C. (ed.), <u>Aluminum, Steel and Organic Reinforced Epoxy Systems</u>, Materials Technology Series, Vol. 6, 1974.

4-47.   Clements, L., p. 28.

4-48.   Critchley, J., Knight, G., and Wright, W., <u>Heat Resistant Polymers</u>, Plenum Press, New York, NY, 1983.

4-49.   Richardson, M., p. 215.

4-50.   Madorsky, S., and Straus, S., <u>Modern Plastic</u>, Volume 38 No. 6, 1961.

4-51.   Critchley, J., p. 28.

4-52.   Critchley, J., p. 80.

4-53.   Economy, J., and Wohrer, L., "Phenolic Fibers," <u>Encyclopedia of Polymer Science and Technology</u>, Interscience Publishers, New York, NY, 1971.

4-54.   Boller, K., "Strength Properties of Reinforced Plastic Laminates at Elevated Temperatures," WADC Technical Report, 1960.

4-55.   Knop, A., and Scheib, W., <u>Chemistry and Application of Phenolic Resins</u>, Springer-Verlag, Berlin, W. Germany, 1979.

4-56.   Doyle, H. and Harrie, S., "Phenolics and Silicones," <u>Handbook of Fiberglass and Advanced Plastics Composites</u>, Van Nostrand Reinhold Co., New York, NY, 1969.

4-57.   Richardson, M., p. 248.

4-58.   Richardson, M., p. 418.

4-59.   Hilado, C., p. 23.

4-60.   Hilado, C., p. 24.

4-61.   Schmidt, P., and Jones, W., "Carbon-Based Fiber Reinforced Plastics," <u>Chemical Engineering Progress</u>, Vol. 58 Number 10, October 1962.

4-62.   Schmidt, P., p. 218.

4-63.   Critchley, J., p. 43.

4-64.   Blais, J., <u>Amino Resins</u>, Reinhold Publishing Co., New York, NY, 1959.

4-65.   Delmonte, J., p. 127.

4-66.   Critchley, J., p. 252.

4-67.   Critchley, J., p. 253.

4-68.   Wright, W., "Application of Thermal Methods to the Study of the
        Degradation of Polyimides," Developments in Polymer Degrada-
        tion, Applied Science Publishers, London, Eng., 1981.

4-69.   Critchley, J., p. 256.

4-70.   Critchley, J., p. 256.

4-71.   Delvigs, P., and Alston, W., "Effect of Graphite Fiber Stability
        on the Properties of PMR Polyimide Composites," SAMPE 24th
        Symposium, Volume 24, May 1979.

4-72.   Jones, J., "Celion/LaRC-160 Graphite/Polyimide Composite Proces-
        sing Techniques and Properties," SAMPE Journal, April 1983.

4-73.   Kunz, S., "Thermodynamical Characterization of Graphite/Polyimide
        Composites," Composites for Extreme Environments, ASTM Pub-
        lication, STP 768, 1982.

4-74.   Kunz, S., p. 37.

4-75.   Campbell, M., and Burleigh, D., "Thermophysical Properties Data
        on Graphite/Polyimide Composite Materials," Composites for
        Extreme Environments, ASTM Publication, STP 768, 1982.

Glossary

1. <u>Aramid (Kevlar) Fiber</u>. Kevlar is a type of polymer fiber such as nylon, rayon, and Nomex. However, the Kevlar fibers possess very high specific strength combined with an attractive specific modulus or stiffness due to the low density of these materials. Kevlar materials have excellent inherent toughness which contributes to the high impact strength. The specific modulus is the highest of the polymer reinforcing fibers and is exceeded in value only by the moduli of advanced ceramic type materials such as graphite, boron, silicon carbide and alumina. The Kevlar materials (Kevlar 29, Kevlar 49) have a negative coefficient of thermal expansion in the longitudinal direction which must be considered in the design and fabrication of composite structures. Kevlar is available in the following reinforcement forms: continuous yarns, rovings, staples, and fabrics.

2. <u>Autoclave Molding</u>. A process that is used primarily for molding thermoset preimpregnated materials.

<u>Process</u>. Prepreg laminates are laid up on a mold duplicating the part surface to be fabricated; the laminate is covered with flexible sheets and vacuum sealed; the assembly is cured under vacuum, heat and pressure according to a specified heating-pressure cycle; as the laminate cures, the material becomes an integrated part duplicating the shape of the mold.

<u>Advantages</u>. (1) Molds are made from inexpensive materials and have minimum wear. (2) Premolded composite and metallic inserts are easily incorporated in the molding process as well as different types of prepreg materials including cloth, mat, and fibers such as glass and graphite in mixed materials. (3) Extremely large parts can be produced in one lay up.

<u>Disadvantages</u>. (1) Autoclave molding is labor intensive because of the hand work involved in laying up the preform and vacuum bagging. (2) The process is slow because of the time required to lay up the part and the time for curing in the autoclave. (3) Only one side of the part is molded since the back is free-formed by bag pressure, and controllability of the tolerance of one face of the part may be unacceptable.

3. <u>Boron Fiber</u>. Synthesized by chemical vapor deposition (CVD) from the reduction of Boron Trichloride (BCl_3) with hydrogen on a tungsten or carbon monofilament substrate. The substrate is resistively heated to a temperature of about 2,300 ^{O}F and continuously pulled through the reactor to obtain the desired boron coating thickness. Current development of boron filament technology is directed at cost reductions for future markets and increased strengths for specialized applications. One area of development directed at lowering filament cost is to replace the tungsten wire precursor with a carbon monofilament. Use of the carbon core offers the potential of reducing the cost of boron filament significantly.

4. <u>Chemical Vapor Deposition (CVD)</u>. A manufacturing process used in coating fibers or parts. Method involves placing the fiber or part to be

coated in a chemical vapor environment. The vapor deposits on the fiber in a thin layer. This method is used to make Borsic filaments. Boron filaments are made, and then by the CVD process, a SiC coating is deposited onto the boron fiber.

5. <u>Composite</u>. A compound material which differs from alloys by retaining the characteristics of individual components which are so incorporated into the composite as to take advantage of their attributes, not their shortcomings. A material which consists of a strengthening phase in the form of particulates, whiskers, short, discontinuous or continuous fibers embedded in another phase called a matrix. Composite materials are usually divided into three broad groups identified by the matrix material: resin, metal and ceramic.

6. <u>Compression Molding</u>. A manufacturing process for fabricating a composite structure. Materials containing the fiber and resin are placed in a metal cavity and compressed under the pressure from a metal punch descending from above. The resulting fluid material is forced into the cavity or mold. After the material has set or cured, the mold is opened and the part is removed.

7. <u>Creep</u>. A slow deformation by stresses below the normal yield strength commonly occurring at elevated temperatures. Creep rate is the creep strain per unit of time.

8. <u>Fatigue</u>. The tendency to fracture under cyclic stresses.

9. <u>Fiberglass Reinforcement</u>. The most common reinforcing fibers for resin matrix composite structures. The two most common are structural glasses and quartz. The tensile strength of "S" glass is higher than that of any other high performance fiber including both boron and graphite fibers.

Two forms of fiberglass can be produced: continuous fiber and staple (discontinuous) fiber. Both forms are made by the same production method up to the fiber drawing stage. Continuous fibers are produced by introducing molten glass consisting of sand, limestone, and boric acid into a fiber drawing furnace. A platinum alloy tank called a bushing allows the molten glass to be gravity-fed through the multiple holes located at the base of the bushing. The droplets of molten glass are gathered together and mechanically attenuated to the proper dimensions, passed through a light water spray, and then traversed over a belt which applies a protective and lubricating binder or size to the individual fibers or filaments.

Staple fibers are produced by passing a jet of air across the holes in the base of the bushing, thus pulling individual fibers eight to fifteen inches long from the molten glass extruding from each opening. These fibers are collected on a rotating drum, sprayed with binder, and gathered into a strand.

10. <u>Filament Winding</u>. Technique which consists of mechanically winding single or multiple continuous fiber strands on a mandrel in a pattern specified by the designer for the required strength or stiffness in the final product. The process consists first of selecting a mandrel, the dimensions of which conform to the inside of the part being formed. The selection of a mandrel depends on the complexity of the part being formed. Typical mandrels are metal cores, inflatable sections, or soluble hard salts. Multiple passes are employed to obtain a desired wall thickness.

<u>Advantages</u>. (1) Ability to place large quantities of composite in a precise position without large expenditure of manpower, (2) Low capital investment, (3) Easy alignment of fibers to carry torsional or circumferential loads efficiently, and (4) Highly automated process.

<u>Disadvantages</u>. (1) Addition of preformed composite or steel inserts may complicate the filament winding process, (2) Longitudinal strength and stiffness properties are low unless angle winding or hand lay up is employed, and (3) Mandrel design fabrication is expensive for complex designs.

11. <u>Graphite Fibers</u>. Fibers consist principally of carbon, which is amorphous, and a lesser amount of graphite, which has a hexagonal crystalline structure. The percentage of graphite in the fibers depends essentially on the final processing temperature. As the final processing temperature is increased, the percentage of graphite is increased. The three most popular precursors are polyacrylonitrile (PAN), pitch, and staple rayon fibers.

12. <u>Injection Molding</u>. A manufacturing technique which produces more thermoplastic products than any other process. Raw material is fed from a hopper into a pressure chamber ahead of a plunger. The plunger forces the plastic into a heating chamber, where it is melted, and its flow is regulated. The plastic then flows through a nozzle seated against the mold and into the die cavity. The die cavity is cooler than the molten plastic so the plastic solidifies as soon as the mold is filled.

<u>Advantages</u>. (1) Parts can be produced at extremely high production rates and at low labor costs, (2) Dimensional tolerances are very precise, and (3) Scrap material is reusable.

<u>Disadvantages</u>. (1) The process is limited to very short fiber reinforcements; therefore, the very high tensile strength and stiffness capabilities of long fiber reinforcement can't be obtained, and (2) Mold costs are high.

13. <u>Matrix</u>. One of the two component materials which makes up a composite. The other is commonly referred to as the reinforcement. It can be a metal, resin, or ceramic material. It holds the reinforcements together to enable the transfer of stresses and loads to the reinforcements.

14. Modulus of Elasticity. The ratio of the change in stress to the change in elastic strain. If the stress-strain relationship is linear, the parameter is called Young's modulus. If the relationship is nonlinear, an instantaneous value of modulus of elasticity can be computed for a specific value of applied stress.

15. Poisson's Ratio. The ratio of material lateral strain to axial strain.

16. Pultrusion. A continuous manufacturing process where resin-coated fiber material is pulled through a heat forming die which results in fully or partially cured parts. The fiber is coated either in a resin dip bath before entering the back end of the die or within the die where the resin has been pressure fed. The resin sets during passage through the die.

Advantages. (1) Capital investment in equipment is low compared to the volume of parts which can be produced, (2) Pultruded parts have high strength and stiffness in the direction of fiber alignment, (3) Labor costs are low, and (4) No scrap loss is incurred.

Disadvantages. (1) Constant cross-section parts are required, and (2) Partially cured pultrusions are more complex and costly.

17. Reinforcement. One of the two component materials which make up a composite. The other is commonly referred to as the matrix. Reinforcement materials come in forms of continuous and discontinuous fibers, whiskers, particulates and wires. Principal reinforcement materials are graphite, boron, glass, and silicon carbide. The material that carries the major stresses and loads.

18. Shear Modulus. The shear stress per unit shear strain.

19. Silicon Carbide Reinforcement. A potential low cost/high performance filament suitable for advanced metal matrix applications. The SiC filament is produced in a manner similar to the boron filament using chemical vapor deposition (CVD) onto continuous substrates. The substrate is resistively heated to incandescent temperatures in a hydrogen/silane gaseous environment to deposit silicon carbide on either a tungsten or carbon substrate. The SiC filament retains strength properties at temperatures well above 1,200 °F with usable strengths at 1,800 to 2,000 °F. Other types of SiC reinforcement are being developed to reduce cost and increase the composite mechanical properties. The development of flat, ribbon shaped filaments and whiskers are alternative forms of SiC. Significant cost reductions and production increases have been seen with the flat, ribbon shape compared to the standard and substrate.

20. Specific Modulus. Defined as E/ρ, where E is the elastic modulus, and ρ is the density.

21. <u>Specific Strength</u>. Defined as σ/ρ, where σ is the tensile strength, and ρ is the density.

22. <u>Thermal Expansion Coefficient</u>. Change in dimensions per change in temperature.